MW01697958

OSTP

Grade 8 Math Practice

GET DIGITAL ACCESS TO

- 2 OSTP Practice Tests
- Personalized Study Plans

REGISTER NOW

Link	QR Code
Visit the link below for online registration **lumoslearning.com/a/tedbooks** **Access Code: OKMQB8**	

Oklahoma School Testing Program Test Prep: 8th Grade Math Practice Workbook and Full-length Online Assessments: OSTP Study Guide

Contributing Editor - Jessica Fisher
Contributing Editor - Wilma Muhammad
Executive Producer - Mukunda Krishnaswamy
Program Director - Anirudh Agarwal
Designer and Illustrator - Sowmya R.

ISBN 13: 979-8897130016

Printed in the United States of America

CONTACT INFORMATION

LUMOS INFORMATION SERVICES, LLC

PO Box 1575, Piscataway, NJ 08855-1575
www.LumosLearning.com

Email: support@lumoslearning.com
Tel: (732) 384-0146
Fax: (866) 283-6471

Lumos Learning
Step Up Your Skills

INTRODUCTION

This book is specifically designed to improve student achievement on the Oklahoma School Testing Program (OSTP). Students perform at their best on standardized tests when they feel comfortable with the test cotent as well as the test format. Lumos online practice tests are meticulously designed to mirror the state assessment. They adhere to the guidelines provided by the state for the number of sessions and questions, standards, difficulty level, question types, test duration and more.

Based on our decade of experience developing practice resources for standardized tests, we've created a dynamic system, the Lumos Smart Test Prep Methodology. It provides students with realistic assessment rehearsal and an efficient pathway to overcoming each proficiency gap.

Use the Lumos Smart Test Prep Methodology to achieve a high score on the OSTP.

Lumos Smart Test Prep Methodology

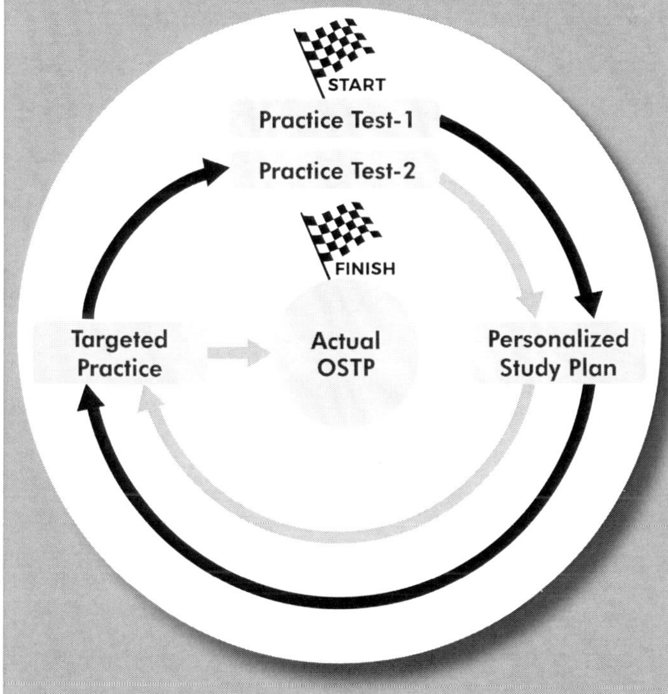

1 The student takes the first online diagnostic test, which assesses proficiency levels in various standards.

2 StepUp generates a personalized online study plan based on the student's performance.

3 The student completes targeted practice in the printed workbook and marks it as complete in the online study plan.

4 The student then attempts the second online practice test.

5 StepUp generates a second individualized online study plan.

6 The student completes the targeted practice and is ready for the actual OSTP.

Table of Contents

Date of Completion:_______________ Score:_______________

Chapter 1
Numbers and Operations

Lesson 1: Properties of Exponents

1. Is -5^2 equal to $(-5)^2$?

Ⓐ Yes, because they both equal -25.
Ⓑ Yes, because they both equal -10.
Ⓒ Yes, because they both equal 25.
Ⓓ No, because -5^2 equals -25 and $(-5)^2$ equals 25.

2. $\frac{X^6}{X^{-2}} =$

Ⓐ $\frac{1}{X^3}$

Ⓑ $\frac{1}{X^{12}}$

Ⓒ X^4

Ⓓ X^8

3. Which of the following is equal to 3^{-2} ?

Ⓐ $\frac{1}{9}$

Ⓑ -9

Ⓒ 9

Ⓓ $\frac{1}{6}$

4. Which of the following is equivalent to $X^{(2-5)}$?

Ⓐ X^3

Ⓑ $X^{\frac{1}{3}}$

Ⓒ $\frac{1}{X^3}$

Ⓓ 3^X

5. $1^9 =$

Ⓐ 1
Ⓑ 3
Ⓒ 9
Ⓓ $\frac{1}{9}$

6. $(X^{-3})(X^{-3}) =$

Ⓐ X^6
Ⓑ X^9
Ⓒ $\frac{1}{X^6}$
Ⓓ $\frac{1}{X^9}$

7. $(X^{-2})^{-7} =$

Ⓐ X^5

Ⓑ X^{14}

Ⓒ $\frac{1}{X^5}$

Ⓓ $\frac{1}{X^{14}}$

8. $(X^4)^0 =$

Ⓐ X
Ⓑ X^4
Ⓒ 1
Ⓓ 0

9. $(3^2)^3 =$

Ⓐ 3^5
Ⓑ 3^6
Ⓒ 3
Ⓓ 1

10. $5^2+5^3 =$ ______

11. Which of the following show the proper laws of exponents?

Note : More than one option may be correct. Select all the correct answers.

Ⓐ $3^2 x 3^5 = 3^{10}$
Ⓑ $(4^2)^3 = 4^6$
Ⓒ $\frac{8^5}{8^1} = 8^4$
Ⓓ $7^4 x 7^4 = 7^8$

12. Simplify this expression.
$a^7(a^8)(a)$

Write your answer in the box below.

13. Select the ones that properly applied the different Laws of Exponents, making sure to keep positive exponents.

Note : More than one option may be correct. Select all the correct answers.

Ⓐ $(4a^3)^2 = 16a^6$
Ⓑ $(2x^4)^2 = 4x^6$
Ⓒ $(x^2y^{-1})^2 = \frac{x^4}{y^2}$
Ⓓ $(2a^{-2})^3 = 8a^6$

Date of Completion:______________ Score:______________

CHAPTER 1 → Lesson 2: Scientific Notation

1. In 2007, approximately 3,380,000 people visited the Statue of Liberty. Express this number in scientific notation.

Ⓐ 0.388×10^7
Ⓑ 3.38×10^6
Ⓒ 33.8×10^5
Ⓓ 338×10^4

2. The average distance from Saturn to the Sun is 890,800,000 miles. Express this number in scientific notation.

Ⓐ 8908×10^8
Ⓑ 8908×10^5
Ⓒ 8.908×10^8
Ⓓ 8.908×10^5

3. The approximate population of Los Angeles is 3.8×10^6 people. Express this number in standard notation.

Ⓐ 380,000
Ⓑ 3,800,000
Ⓒ 38,000,000
Ⓓ 380,000,000

4. The approximate population of Kazakhstan is 1.53×10^7 people. Express this number in standard notation.

Ⓐ 153,000
Ⓑ 1,530,000
Ⓒ 15,300,000
Ⓓ 153,000,000

5. The typical human body contains about 2.5×10^{-3} kilograms of zinc. Express this amount in standard form.

Ⓐ 0.00025 kilograms
Ⓑ 0.0025 kilograms
Ⓒ 0.025 kilograms
Ⓓ 0.25 kilograms

6. If a number expressed in scientific notation is N x 10^5, how large is the number?

Ⓐ Between 1,000 (included) and 10,000
Ⓑ Between 10,000 (included) and 100,000
Ⓒ Between 100,000 (included) and 1,000,000
Ⓓ Between 1,000,000 (included) and 10,000,000

7. Red light has a wavelength of 650 x 10^{-9} meters. Express the wavelength in scientific notation.

Ⓐ 65.0 x 10^{-10} meters
Ⓑ 65.0 x 10^{-8} meters
Ⓒ 6.50 x 10^{-7} meters
Ⓓ 6.50 x 10^{-11} meters

8. A strand of hair from a human head is approximately 1 x 10^{-4} meters thick. What fraction of a meter is this?

Ⓐ $\frac{1}{100}$
Ⓑ $\frac{1}{1,000}$
Ⓒ $\frac{1}{10,000}$
Ⓓ $\frac{1}{100,000}$

9. Which of the following numbers has the greatest value?

Ⓐ 8.93 x 10^3
Ⓑ 8.935 x 10^2
Ⓒ 8.935 x 10^3
Ⓓ 893.5 x 10^1

10. Which of the following numbers has the least value?

Ⓐ -1.56 x 10^2
Ⓑ -1.56 x 10^3
Ⓒ 1.56 x 10^2
Ⓓ 1.56 x 10^3

11. Which of the following are correctly written in scientific notation?

Note that more than one option may be correct. Select all the correct options

Ⓐ $.032 \times 10^5$
Ⓑ 11.002×10^{-1}
Ⓒ 1.23×10^5
Ⓓ 9.625×10^{-7}

12. Change 2,347,000,000 from standard form to scientific notation by filling in the blank boxes.

$\square \times 10^{\square}$

13. Convert 0.0000687 to scientific notation by filling in the blank boxes.

$\square \times 10^{\square}$

CHAPTER 1 → Lesson 3: Solving Problems Involving Scientific Notation

1. The population of California is approximately 3.7×10^7 people. The land area of California is approximately 1.6×10^5 square miles. Divide the population by the area to find the best estimate of the number of people per square mile in California.

Ⓐ 24 people
Ⓑ 240 people
Ⓒ 2,400 people
Ⓓ 24,000 people

2. Mercury is approximately 6×10^7 kilometers from the Sun. The speed of light is approximately 3×10^5 kilometers per second. Divide the distance by the speed of light to determine the approximate number of seconds it takes light to travel from the Sun to Mercury.

Ⓐ 2 seconds
Ⓑ 20 seconds
Ⓒ 200 seconds
Ⓓ 2,000 seconds

3. Simplify $(4 \times 10^6) \times (2 \times 10^3)$ and express the result in scientific notation.

Ⓐ 8×10^9
Ⓑ 8×10^{18}
Ⓒ 6×10^9
Ⓓ 6×10^{18}

4. Simplify $(2 \times 10^{-3}) \times (3 \times 10^5)$ and express the result in scientific notation.

Ⓐ 5×10^{-8}
Ⓑ 5×10^{-15}
Ⓒ 6×10^8
Ⓓ 6×10^2

5. Washington is approximately 2.4×10^3 miles from Utah. Mary drives 6×10^1 miles per hour from Washington to Utah. Divide the distance by the speed to determine the approximate number of hours it takes Mary to travel from Washington to Utah.

Ⓐ 41 hours
Ⓑ 40 hours
Ⓒ 39 hours
Ⓓ 38 hours

6. Which of the following is NOT equal to $(5 \times 10^5) \times (9 \times 10^{-3})$?

Ⓐ 4.5×10^4
Ⓑ 4.5×10^3
Ⓒ 4,500
Ⓓ 45 x 100

7. Find $(5 \times 10^7) \div (10 \times 10^2)$ and express the result in scientific notation.

Ⓐ 5×10^4
Ⓑ 0.5×10^5
Ⓒ 50×10^9
Ⓓ 5.0×10^9

8. Approximate .00004567 x .00001234 and express the result in scientific notation.

Ⓐ 5.636×10^{-8}
Ⓑ 5.636×10^{-9}
Ⓒ 5.636×10^{-10}
Ⓓ None of the above.

9. Find the product $(50.67 \times 10^4) \times (12.9 \times 10^3)$ and express the answer in standard notation.

Ⓐ 653.643
Ⓑ 65,364,300,000
Ⓒ 6.53643×10^9
Ⓓ 6,536,430,000

10 Approximate the quotient and express the answer in standard notation.

$(1.298 \times 10^4) \div (3.97 \times 10^2)$

Ⓐ 32.7
Ⓑ .327
Ⓒ $.327 \times 10^2$
Ⓓ None of the above.

11. Which of the following examples correctly apply the rules of scientific notation? Select all the correct answers.

Ⓐ $(4.0 \times 10^3)(5.0 \times 10^5) = 2 \times 10^9$

Ⓑ $\dfrac{4.5\times10^5}{9.0\text{x}10^9} = 2 \times10^4$

Ⓒ $(2.1 \times 10^5) + (2.7 \times 10^5) = 4.8 \times 10^5$

Ⓓ $(3.1 \times 10^5) - (2.7 \times 10^2) = 0.4 \times 10^3$

12. Which of the following lists are ordered from greatest to least? Select all the correct answers.

Ⓐ 2.0×10^2, 3.0×10^6, 4.0×10^{-7}, 5.0×10^{12}

Ⓑ 4.0×10^{-7}, 2.0×10^2, 3.0×10^6, 5.0×10^{12}

Ⓒ 3.0×10^7, 3.0×10^6, 3.0×10^2, 3.0×10^{-7}

Ⓓ 1.8×10^9, 1.5×10^6, 1.4×10^{-5}, 1.9×10^{-8}

13. $(6\text{x}10^3)(9.91\text{x}10^0)$ =___________

Date of Completion:_______________ Score:_______________

CHAPTER 1 → Lesson 4: Square & Cube Roots

1. What is the cube root of 1,000 ?

Ⓐ 10

Ⓑ 100

Ⓒ $33\frac{1}{3}$

Ⓓ $333\frac{1}{3}$

2. $8\sqrt{12} \div \sqrt{15} =$

Ⓐ $\frac{4}{5}$

Ⓑ $\frac{8}{5}$

Ⓒ $\frac{16}{\sqrt{5}}$

Ⓓ $\frac{\sqrt{5}}{8}$

3. The square root of 75 is between which two integers?

Ⓐ 8 and 9
Ⓑ 7 and 8
Ⓒ 9 and 10
Ⓓ 6 and 7

4. The square root of 110 is between which two integers?

Ⓐ 10 and 11
Ⓑ 9 and 10
Ⓒ 11 and 12
Ⓓ 8 and 9

5. Solve the following problem: $6\sqrt{20} \div \sqrt{5} =$ _____

Ⓐ 12
Ⓑ 11
Ⓒ 30
Ⓓ 5

6. The cube root of 66 is between which two integers?

Ⓐ 4 and 5
Ⓑ 3 and 4
Ⓒ 5 and 6
Ⓓ 6 and 7

7. Which expression has the same value as $3\sqrt{144} \div \sqrt{12}$?

Ⓐ $3\sqrt{12}$
Ⓑ $4\sqrt{12}$
Ⓒ $27 \div \sqrt{12}$
Ⓓ $33 \div \sqrt{12}$

8. The cubic root of 400 lies between which two numbers?

Ⓐ 5 and 6
Ⓑ 6 and 7
Ⓒ 7 and 8
Ⓓ 8 and 9

9. Which of the following is equivalent to the expression $4\sqrt{250} \div 5\sqrt{2}$?

Ⓐ $4\sqrt{25} \div 5$
Ⓑ $4\sqrt{125} \div \sqrt{2}$
Ⓒ $4\sqrt{10}$
Ⓓ $4\sqrt{5}$

10. The cube root of 150 is closest to which of the following?

Ⓐ 15
Ⓑ 10
Ⓒ 5
Ⓓ 3

11. Select all that apply: What is $\sqrt{\frac{81}{289}}$?

Ⓐ $\frac{1}{2}$

Ⓑ $\frac{9}{17}$

Ⓒ $\frac{-1}{2}$

Ⓓ $\frac{-9}{17}$

12. Select all the numbers which have integers as the cube roots.

Ⓐ $\sqrt[3]{27}$
Ⓑ $\sqrt[3]{9}$
Ⓒ $\sqrt[3]{1000}$
Ⓓ $\sqrt[3]{18}$

13. Fill in the boxes to make the statement true

$\sqrt[3]{8} = \square$ since $\square \times \square \times \square = 8$

End of Numbers & Operations

ANSWER KEY AND DETAILED EXPLANATION

Chapter 1 Numbers & Operations

Lesson 1: Properties of Exponents

Question No.	Answer	Detailed Explanation
1	D	Unless there are parentheses to denote otherwise, the exponent is only applied to the constant or variable immediately preceding it. If there are parentheses immediately preceding it, then it is applied to everything within the parentheses.
2	D	When dividing quantities with like bases, you must subtract the exponents. 6 - (-2) =8 $\frac{X^6}{X^{-2}} = X^{6-(-2)} = X^8$
3	A	A quick way to change exponents from negative to positive is to move the expression to which the negative exponent is applied from the denominator to the numerator or vice versa and change the sign of the exponent. 3^{-2} is the same as $\frac{1}{3^2}$ which is the same as $\frac{1}{9}$.
4	C	$X^{(2-5)} = X^{-3}$ Now move x^{-3} to the denominator and change the sign of the exponent from negative to positive. $\frac{1}{X^3}$
5	A	Regardless of the number of 1s that we multiply the result is always 1 because 1 is the identity element for multiplication.
6	C	When multiplying quantities with the same base, you add exponents. $(X^{-3})(X^{-3}) = X^{-6}$ To change the exponent -6 to positive 6, you write the reciprocal of X^{-6}. $\frac{1}{X^6}$
7	B	$(x^{-2})^{-7} = x^{14}$ because to raise a power to a power, we multiply exponents.

Question No.	Answer	Detailed Explanation
8	C	4 x 0 = 0 and any number (other than 0, as 0^0 is not defined) to the 0 power is 1 by definition.
9	B	When raising a power to a power, multiply exponents.
10	150	$5^2 = 25$ and $5^3 = 125$ $25 + 125 = 150$
11	B,C & D	Option B is using the Power of a Power Property $(4^2)^3 = 4^6$ because $4^2 \times 4^2\ 4^2 = 4^{2+2+2} = 4^6$. Option C is using Division Property of Exponents $8^5/8^1 = 8^4$. This is true because $\frac{8\times8\times8\times8\times8}{8}$ when simplified, leaves you with $8\times8\times8\times8$ which is 8^4. Option D is using the Multiplicative Property of Exponents. $7^4 \times 7^4 = 7^8$ because $7\times7\times7\times7\times7\times7\times7\times7 = 7^8$
12	a^{16}	You are using your product rule to find out the exponent. When using the product rule if the base is the same, then you add the exponents. In this case, you will add 7+8+1=16. So it would become a^{16}.
13	A, C	In the expression $(4a^3)^2$ you are applying the Power of a Power Law. You distribute the squared to each term in the parentheses to end up with 4^2a^{3x2} which simplifies to $16a^6$. The same thing applies to $(x^2y^{-1})^2$ so you end up with x^{2x2} and y^{-1x2}. These simplify to x^4 and y^{-2}. In this case, since you want to keep positive exponents, you have to move the y term to the denominator to keep positive exponents. Thus making it $\frac{x^4}{y^2}$

Lesson 2: Scientific Notation

Question No.	Answer	Detailed Explanation
1	B	Moving the decimal in 3,380,000 just to the right of the first non-zero digit requires us to move it 6 places to the left resulting in 3.38×10^6.
2	C	Moving the decimal in 890,800,000 just to the right of the first non-zero digit requires us to move it 8 places to the left resulting in 8.908×10^8.
3	B	Moving the decimal to the right 6 places from 3.8×10^6, we get 3,800,000.
4	C	From 1.53×10^7, we move the decimal to the right 7 places giving us 15,300,000.
5	B	Move the decimal 3 places to the left.
6	C	$1 \leq N \leq 9.99...$ and we must move the decimal 5 places to the right resulting in $100{,}000 \leq N \times 10^5 < 1{,}000{,}000$.
7	C	To change to standard notation from 650×10^{-9}, We must move the decimal 9 places to the left because the exponent is negative. The result is 0.000000650. Changing to scientific notation, we get 6.5×10^{-7}. Or, since we must move the decimal 2 places to the left to get it properly placed for scientific notation, we only need to move it 7 more places for standard notation; so scientific notation would be written as 6.50×10^{-7}. Alternate method : $650 \times 10^{-9} = 6.50 \times 10^2 \times 10^{-9}$. Using laws of exponents, we get $10^2 \times 10^{-9} = 10^{-7}$. Therefore, wavelength of red light in scientific notation is 6.50×10^{-7} m
8	C	$1 \times 10^4 = .0001$ (By moving decimal 4 places to the left because the exponent is -4. Since the 1 is in the ten-thousandths place, the fraction will be $\frac{1}{10{,}000}$
9	C, D	$8.935 \times 10^3 = 8{,}935$ which is the greatest of the numbers shown.
10	B	Since we have negative numbers and positive numbers, the lowest value will be negative. -1,560 is less than -156.

Question No.	Answer	Detailed Explanation
11	C, D	Both the options (C) and (D) are correct because their first number is a decimal that is greater than or equal to 1 but less than 10. They are then multiplied by a power of 10. One is multiplied by a positive power of 10, making it a large number and the second one is multiplied by a negative power making it a really small number. Both are in correct form.
12	2.347×10^9	The number in the first box has to be a decimal number that is greater than or equal to 1, but less than 10. In this case it will be 2.347. Next, we will take it 10 times itself to an exponent which will move our decimal until it becomes our number in standard form. So in this case we have to take 10 times itself 9 times, so our exponent will be 9. Our final answer will be 2.347×10^9
13	6.87×10^{-5}	This time we are changing a really small number into scientific notation. The same principles apply. First we move our decimal to make it a number greater than one, but less than 10. In this case it becomes 6.87. Next we figure out the power of ten we need to move the decimal over that far. Since we are moving the decimal in the opposite direction, our power of ten will be negative this time. Here we are moving it 5 places, so it is 10^{-5}. Combine the two parts and you get 6.87×10^{-5}.

Lesson 3: Solving Problems Involving Scientific Notation

Question No.	Answer	Detailed Explanation
1	B	3.7×10^7 people ÷ 1.6×10^5 square miles =(3.7 ÷ 1.6) x 10^{7-5} = 2.3125×10^2 people/square mile = 231.25 people/ sq. mi. 240 is the best estimate.
2	C	6×10^7 kilometers ÷ 3×10^5 kilometers per second = (6÷3) x $10^{(7-5)}$ = 2×10^2(kilometers ÷ kilometers/seconds) = 2×10^2 seconds = 200 seconds
3	A	$(4 \times 10^6) \times (2 \times 10^3) = 8 \times 10^{(6+3)} = 8 \times 10^9$
4	D	$(2 \times 10^{-3}) \times (3 \times 10^5) = (2x3) \times 10^{(-3+5)} = 6 \times 10^2$
5	B	$(2.4 \times 10^3) / (6 \times 10) = 0.4 \times 10^2 = 4x10 = 40$
6	A	$(5x10^5) \times (9x10^{-3})=(45x10^{5-3})=45x10^2=4.5 \times 10^3$ Therefore, 4.5×10^4 is not equal to 4.5×10^3, so 4.5×10^4 is the correct answer. Options (B), (C) and (D) are different ways of expressing the same number.
7	A	$(5 \times 10^7) \div (10 \times 10^2) = 0.5 \times 10^{(7-2)} = 0.5 \times 10^5 = 5 \times 10^4$
8	C	.00004567 x .00001234 = $(4.567 \times 10^{-5}) \times (1.234 \times 10^{-5})$ = $(4.567 \times 1.234) \times 10^{-5+(-5)} = 5.635678 \times 10^{-10}$ $5.636x10^{-10}$ is the correct answer.
9	D	$(50.67 \times 10^4) \times (12.9 \times 10^3) = 653.643 \times 10^{(4+3)} = 653.643 \times 10^7$ = 6,536,430,000 6,536,430,000 is the correct answer.
10	A	$(1.298 \times 10^4) \div (3.97 \times 10^2) = 0.326952 \times 10^2 = 32.7$ 32.7 is the best answer.
11	A, C	$(4.0 \times 10^3)(5.0 \times 10^5) = (4 \times 5) \times (10^{3+5}) = 20 \times 10^8 = 2.0 \times 10^9$. Therefore option (A) is correct. $(4.5 \times 10^5) / (9.0 \times 10^9) = (45 \times 10^4)/(9 \times 10^9) = (45/9) \times 10^{4-9} = 5.0 \times 10^{-5}$. This is not equal to 2.0×10^4. Therefore, option (B) is incorrect. $(2.1 \times 10^5) + (2.7 \times 10^5) = (2.1 + 2.7) \times 10^5 = 4.8 \times 10^5$. Therefore option (C) is correct. $(3.1 \times 10^5) - (2.7 \times 10^2) = (3.1 \times 10^3 - 2.7) \times 10^2 = (3100 - 2.7) \times 10^2 = (3097.3) \times 10^2 = 3.097 \times 10^5$. This is not equal to 0.4×10^3. Therefore option (D) is incorrect.

Question No.	Answer	Detailed Explanation
12	C, D	In these particular cases it is easy to see which numbers are larger and which are smaller based only on their exponents. If it is a larger number the exponent will be greater. So to get them in order from largest to smallest, look at the exponent and put them in order from largest exponent to the smallest, making sure that when you get into negative powers (or exponents), larger the absolute value of the power, smaller the actual number.
13	5.946×10^4	When multiplying scientific notation, you multiply the numbers out front first. In this case we end up with 59.46. Then since we are multiplying and the bases are the same, we add our exponents. In this case we end up with 10^3. However, since our decimal is greater than 10, we have to move our decimal one place to the left. Thus, multiplying it by another power of 10. We need to add that power of 10 to our original 10^3. So our final answer is 5.946×10^4.

Lesson 4: Square & Cube Roots

Question No.	Answer	Detailed Explanation
1	A	$10^3 = 1{,}000$. So, the cube root of 1,000 is 10.
2	C	$8\sqrt{12} = 8\sqrt{(4 \times 3)} = 8(2)\sqrt{3} = 16\sqrt{3}$ $\sqrt{15} = \sqrt{(5 \times 3)}$ So $8\sqrt{12} \div \sqrt{15} = 16\sqrt{3} \div \sqrt{5}\sqrt{3}$
3	A	$8^2 = 64$ and $9^2 = 81$ Therefore, the square root of 75 is between 8 and 9.
4	A	$10^2 = 100$ and $11^2 = 121$ Therefore, the square root of 110 is between 10 and 11.
5	A	$6\sqrt{20} \div \sqrt{5} = 6\sqrt{4}\sqrt{5} \div \sqrt{5} = 6\sqrt{4} = 12$
6	A	$4^3 = 64$ and $5^3 = 125$ Therefore, the cube root of 66 is between 4 and 5.
7	A	$3\sqrt{144} \div \sqrt{12} = 3(12) \div \sqrt{12} = (3 \times \sqrt{12} \times \sqrt{12}) \div \sqrt{12} = 3\sqrt{12}$
8	C	$7^3 = 343$ and $8^3=512$ Therefore, the cube root of 400 lies between 7 and 8.
9	D	$4\sqrt{250} \div 5\sqrt{2} = 4\sqrt{25}\sqrt{10} \div 5\sqrt{2} = 4(5)\sqrt{10} \div 5\sqrt{2} = 4\sqrt{10} \div \sqrt{2} = 4\sqrt{5}\sqrt{2} \div \sqrt{2} = 4\sqrt{5}$
10	C	$15^3 = 3{,}375$; $10^3 = 1{,}000$; $5^3 = 125$; $3^3 = 27$ 150 is closest to 5^3 or 125.
11	B, D	Both the options (B) and (D) are correct because if you take the square root of 81 it can be either -9 or 9. The same applies to fractions. When you take the square root of both the numerator and the denominator you have to consider that the factors can be either positive or negative.
12	A, C	Options (A) and (C) are correct, because if you take the cube root of them, you will get an integer. 27 can be written as $3 \times 3 \times 3 = 3^3$ and 1000 can be written as $10 \times 10 \times 10=10^3$. $\sqrt[3]{27} = 3$ $\sqrt[3]{1000} = 10$ 27, 1000 are called perfect cubes.
13	$\sqrt[3]{8}=2$ since $2 \times 2 \times 2 = 8$	Taking the cube root of a number is finding a number that multiplied by itself 3 times will give you the number under the radical. In this case the cube root of 8 will equal 2 because $2 \times 2 \times 2 = 8$.

Date of Completion:______________ Score:______________

Chapter 2
Algebraic Reasoning & Algebra

Lesson 1: Functions

1. Which of the following is NOT a function?

Ⓐ {(2, 3), (4, 7), (8, 6)}
Ⓑ {(2, 2), (4, 4), (8, 8)}
Ⓒ {(2, 3), (4, 3), (8, 3)}
Ⓓ {(2, 3), (2, 7), (8, 6)}

2. Which of the following tables shows that y is a function of x?

Ⓐ

x	y
1	4
1	7
4	7

Ⓑ

x	y
1	7
3	8
4	7

Ⓒ

x	y
3	2
3	7
4	8

Ⓓ

x	y
1	7
4	7
4	9

3. If y is a function of x, which of the following CANNOT be true?

Ⓐ A particular x value is associated with two different y values.
Ⓑ Two different x values are associated with the same y value.
Ⓒ Every x value is associated with the same y value.
Ⓓ Every x value is associated with a different y value.

4. Each of the following graphs consists of two points. Which graph could NOT represent a function?

Ⓐ 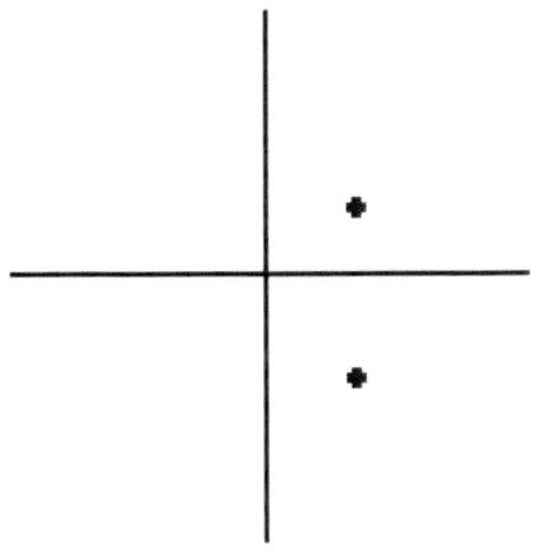

Ⓑ 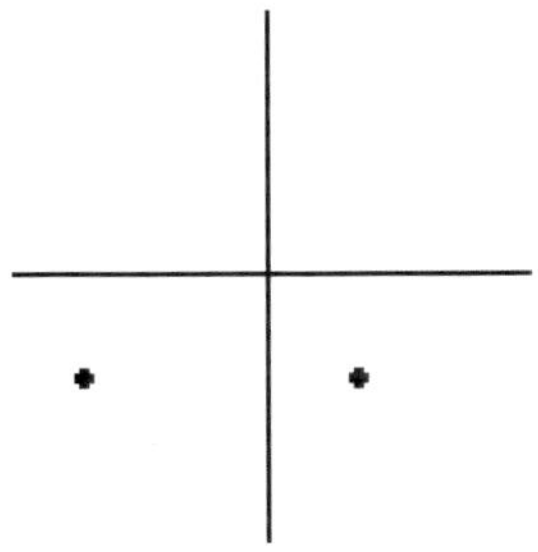

Ⓒ 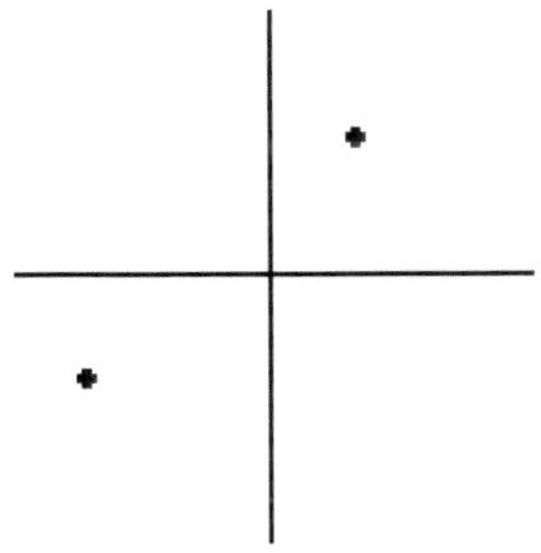

Ⓓ 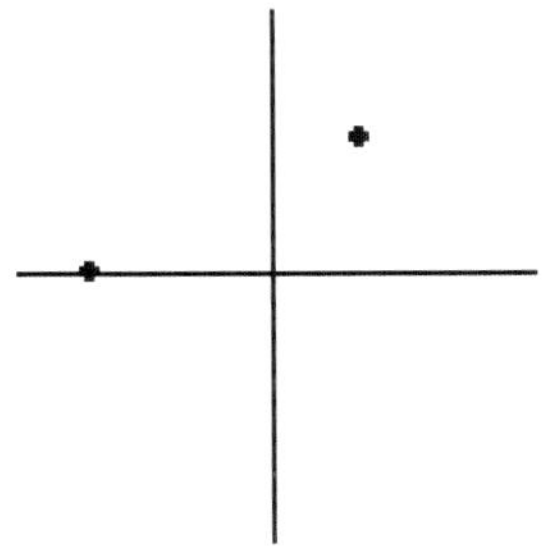

5. Which of the following could NOT be the graph of a function?

Ⓐ 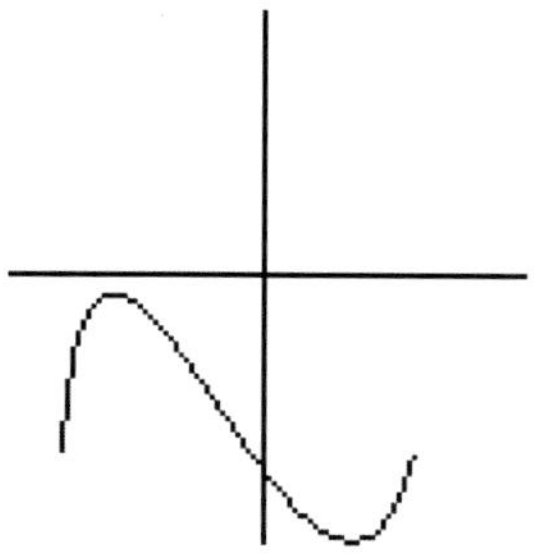

Ⓑ 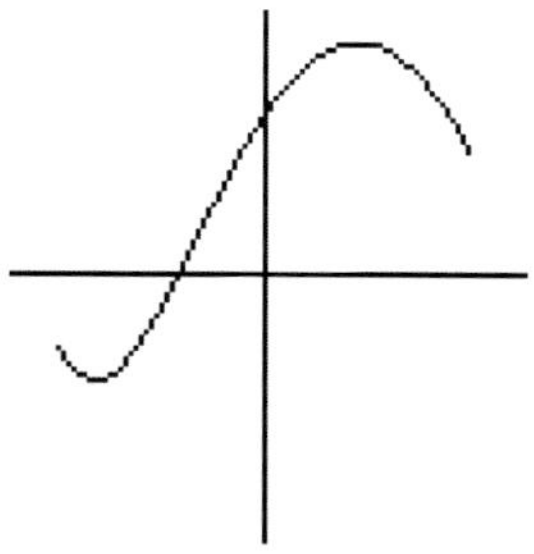

Ⓒ 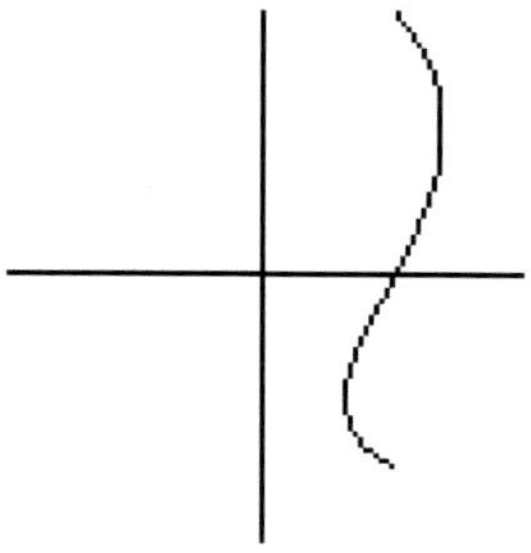

Ⓓ 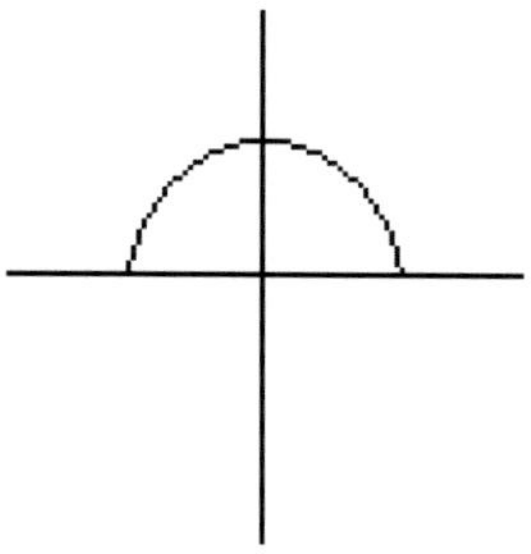

6. Which of the following is NOT a function?

Ⓐ {(0, 0), (2, 2), (4, 4)}
Ⓑ {(0, 4), (2, 4), (4, 4)}
Ⓒ {(0, 0), (2, 0), (4, 0)}
Ⓓ {(0, 0), (0, 2), (0, 4)}

7. Which of the following is true of the graph of any non-constant function?

Ⓐ A line drawn parallel to the x-axis will never cross the graph.
Ⓑ A line drawn perpendicular to the x-axis will never cross the graph.
Ⓒ A line drawn through the graph parallel to the x-axis will cross the graph one and only one time.
Ⓓ A line drawn through the graph perpendicular to the x-axis will cross the graph one and only one time.

8. The given set represents a function:
{(0,1), (1,1), (2,1)}
If the ordered pair ____ was added to the set, it would no longer be a function.

Ⓐ (3,1)
Ⓑ (3,2)
Ⓒ (3,3)
Ⓓ (2,3)

9. In order for this set, {(6,5), (5, 4), (4,3)}, to remain a function, which of the following ordered pairs COULD be added to it?

Ⓐ (6,6)
Ⓑ (5,5)
Ⓒ (4,4)
Ⓓ (3,3)

10. Which of the following sets of ordered pairs represents a function?

Ⓐ [(0,1), (0,2), (1,3), (1,4)]
Ⓑ [(1,1), (1,2), (1,3), (1,4)]
Ⓒ [(2,5), (2,6), (4,7), (5,7)]
Ⓓ [(-7,10), (7,10), (8,9), (9,10)]

11. Select the sets of ordered pairs that represent a function.

There can be more than one correct option. Select all the correct options.

Ⓐ {(1,1),(2,2),(3,3),(4,4)}
Ⓑ {(1,-2),(-2,0),(-1,2),(1,3)}
Ⓒ {(-2,3),(0,1),(2,-1),(3,-2)}
Ⓓ {(-1,7),(0,-3),(1,10),(0,7)}

12. Write the function that would go with the table.

x	2	3	4	5
y	-1	0	1	2

13. Select whether or not each example represents a function.

	Function	Not a Function
{(2,10),(2,20),(4,20),(6,30)}	○	○
{(4,10),(8,12),(4,11),(12,13)}	○	○
{(1,3),(2,4),(3,7),(4,13)}	○	○
y=7x-2	○	○

Date of Completion:_______________ Score:_______________

CHAPTER 2 → Lesson 2: Linear Function Models

1. If a graph includes the points (2, 5) and (8, 5), which of the following must be true?

Ⓐ It is the graph of a linear function.
Ⓑ It is the graph of an increasing function.
Ⓒ It is not the graph of a function.
Ⓓ None of the above

2. The graph of a linear function y = mx + 2 goes through the point (4, 0). Which of the following must be true?

Ⓐ m is negative.
Ⓑ m = 0
Ⓒ m is positive
Ⓓ Cannot be determined.

3. The graph of a linear function y = 2x + b passes through the point (-5, 0). Which of the following must be true?

Ⓐ b is positive.
Ⓑ b is negative.
Ⓒ b = 0
Ⓓ Cannot be determined.

4. The graph of a linear function y = mx + b goes through the point (0, 0). Which of the following must be true?

Ⓐ m is positive.
Ⓑ m is negative.
Ⓒ m = 0
Ⓓ b = 0

5. The graph of a linear function y = mx + b includes the points (2, 5) and (9, 5). Which of the following gives the correct values of m and b?

Ⓐ m = 5 and b = 7
Ⓑ m = 1 and b = 0
Ⓒ m = 0 and b = 5
Ⓓ m = 2 and b = 9

6. What is the slope of the linear function represented by the (x, y) pairs shown in the table below?

x	y
0	11
2	7
3	5

Ⓐ $-\frac{1}{2}$
Ⓑ $\frac{1}{2}$
Ⓒ - 2
Ⓓ 2

7. The graph of a certain linear function includes the points (-4, 1) and (5, 1). Which of the following statements describes the function accurately?

Ⓐ It is an increasing linear function.
Ⓑ It is a decreasing linear function.
Ⓒ It is a constant function.
Ⓓ It is a nonlinear function.

8. Which of the following linear functions has the greatest slope?

Ⓐ

x	y
0	1
2	5
4	9

Ⓑ

x	y
0	3
2	6
4	9

Ⓒ

x	y
0	5
2	7
4	9

Ⓓ

x	y
0	7
2	8
4	9

9. A young child is building a tower of blocks on top of a bench. The bench is 18 inches high, and each block is 3 inches high. Which of the following functions correctly relates the total height of the tower (including the bench) h, in inches, to the number of blocks b?

Ⓐ $h = 3b - 18$
Ⓑ $h = 3b + 18$
Ⓒ $h = 18b - 3$
Ⓓ $h = 18b + 3$

10. Jim owes his parents $10. Each week, his parents pay him $5 for doing chores. Assuming that Jim does not earn money from any other source and does not spend any of his money. Which of the following functions correctly relates the total amount of money m, in dollars, that Jim will have to the number of weeks w?

Ⓐ $m = -5w - 10$
Ⓑ $m = -5w + 10$
Ⓒ $m = 5w - 10$
Ⓓ $m = 5w + 10$

11. An amusement park charges $5 admission and an additional $2 per ride. Which of the following functions correctly relates the total amount paid p, in dollars, to the number of rides, r?

Ⓐ $p = 2r + 5$
Ⓑ $p = 5r + 2$
Ⓒ $p = 10r$
Ⓓ $p = 7r$

12. Which of the following linear functions has the smallest slope?

Ⓐ

x	y
1	2
4	6
7	8

Ⓑ

x	y
1	3
4	6
7	9

Ⓒ

x	y
1	4
4	6
7	8

Ⓓ

x	y
1	5
4	6
7	7

13. The graph of a linear function includes the points (6, 2) and (9, 4). What is the y intercept of the graph?

Ⓐ (0, 0)
Ⓑ (0, -2)
Ⓒ (3, 0)
Ⓓ (0, 3)

14. Which of the following functions has this set of points as solutions?
{(0, -5), (1, 0), and (4, 15)}

Ⓐ $f(x) = 4x - 15$
Ⓑ $f(x) = 0$
Ⓒ $f(x) = -5$
Ⓓ $f(x) = 5x - 5$

15. A music store is offering a special on CDs. The cost is $20.00 for the first CD and $10.00 for each additional CD purchased.
Which of the following functions represents the total amount in dollars of your purchase where x is the number of CDs purchased?

Ⓐ $f(x) = 10x + 20$
Ⓑ $f(x) = 10(x - 1) + 20$
Ⓒ $f(x) = 20x + 10$
Ⓓ $f(x) = 20(x - 1) + 10$

16. A trainer for a professional football team keeps track of the amount of water players consume throughout practice. The trainer observes that the amount of water consumed is a linear function of the temperature on a given day. The trainer finds that when it is 90°F the players consume about 220 gallons of water, and when it is 76°F the players consume about 178 gallons of water. Fill in the value of m in the function : $y = mx - 50$.

y = _____ x - 50

17. Match each slope and coordinates of point with the correct equation

	$y=-4x-3$	$y=7x-5$	$y=-x+2$
(1, 2) slope=7	○	○	○
(-2, 5) slope=-4	○	○	○
(3, -1) slope=-1	○	○	○

18. A cell phone company charges $89.99 for a new phone and then $19.99 per month. What is the slope or rate of change for this situation?

Slope = ________

19. Plot the (x,y) points shown in the below table on a graph such that it represents a linear function.

x	y
0	11
2	7
3	5

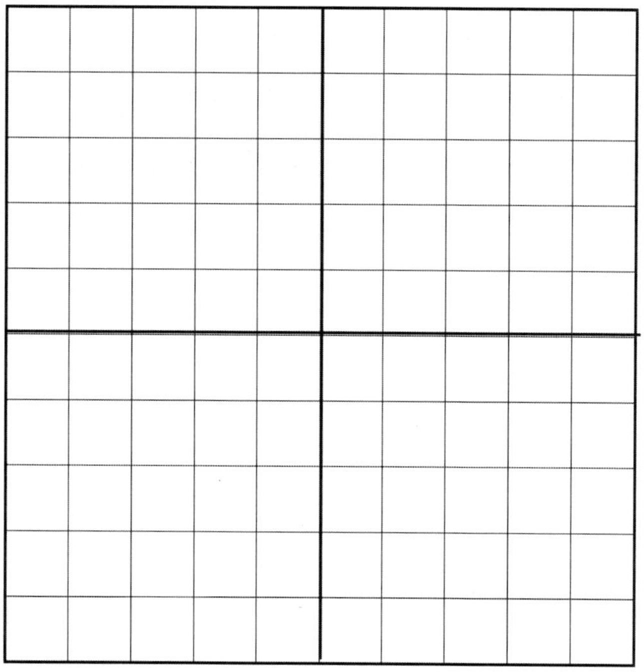

CHAPTER 2 → Lesson 3: Linear Functions

1. A linear function includes the ordered pairs (2, 5), (6, 7), and (k, 11). What is the value of k?

Ⓐ 8
Ⓑ 10
Ⓒ 12
Ⓓ 14

2. Which of the following functions is NOT linear?

Ⓐ $f(x) = x + 0.5$
Ⓑ $f(x) = -x + 0.5$
Ⓒ $f(x) = x^2 + 0.5$
Ⓓ $f(x) = 0.5x$

3. Which of the following functions is linear and includes the point (3, 0)?

Ⓐ $f(x) = 3/x$
Ⓑ $f(x) = x - 3$
Ⓒ $f(x) = 3$
Ⓓ $f(x) = 3x$

4. Four (x, y) pairs of a certain function are shown in the table below. Which of the following describes the function?

x	y
-3	1
-1	4
1	7
3	10

Ⓐ The function increases linearly.
Ⓑ The function decreases linearly.
Ⓒ The function is constant.
Ⓓ The function is not linear.

5. Four (x, y) pairs of a certain function are shown in the table below. Which of the following statements describes the function correctly?

x	y
0	3
1	4
2	7
3	12

Ⓐ The function is linear because it does not include the point (0, 0).
Ⓑ The function is linear because it does not have the same slope between different pairs of points.
Ⓒ The function is nonlinear because it does not include the point (0, 0).
Ⓓ The function is nonlinear because it does not have the same slope between different pairs of points.

6. The graph of a linear function lies in the first and fourth quadrants. Which of the following CANNOT be true?

Ⓐ It is an increasing function.
Ⓑ It is a constant function.
Ⓒ It also lies in the second quadrant.
Ⓓ It also lies in the third quadrant.

7. A linear function includes the ordered pairs (0, 1), (3, 3), and (9, n). What is the value of n?

Ⓐ 5
Ⓑ 6
Ⓒ 7
Ⓓ 8

8. A linear function includes the ordered pairs (0, 3), (3, 9), and (9, n). What is the value of n?

Ⓐ 20
Ⓑ 14
Ⓒ 21
Ⓓ 22

9. The graph of a linear function with a non-negative slope lies in the first and second quadrants. Which of the following CANNOT be true?

Ⓐ It is an increasing function.
Ⓑ It is a constant function.
Ⓒ It also lies in the third quadrant.
Ⓓ It also lies in the fourth quadrant.

10. Which function is represented by this table?

x	y
0	2
1	4
2	6
3	8

Ⓐ $y = 2x + 2$
Ⓑ $y = 3x - 4$
Ⓒ $y = 4x - 5$
Ⓓ $y = 6x - 8$

11. Which of the following functions are linear? Select all the correct answers.

Ⓐ 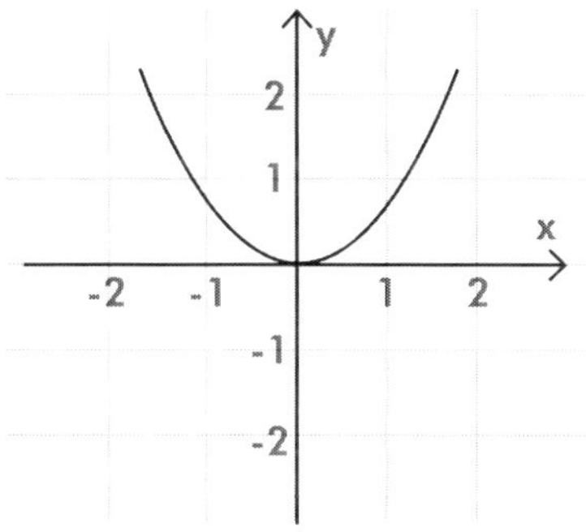

Ⓒ

x	y
-2	-4
-1	-1
0	2
1	5
2	8

difference of y-values

$-1 + 4 = 3$

$2 + 1 = 3$

$5 - 2 = 3$

$8 - 5 = 3$

Ⓑ $y = 5x+9$

Ⓓ $y=x^2+6$

12. **Write whether this represents a linear or non-linear function.**

$2x^2+3y=10$

13. **Match each function to linear or non-linear.**

<table>
<tr><th></th><th>Linear</th><th>Non-Linear</th></tr>
<tr><td>$y = x^2$</td><td>○</td><td>○</td></tr>
<tr><td>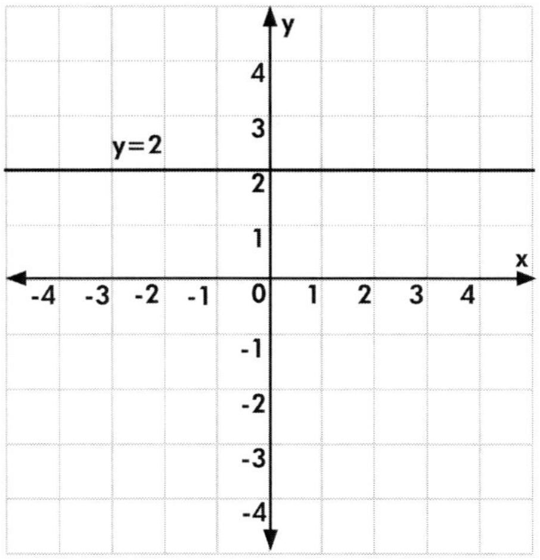
</td><td>○</td><td>○</td></tr>
<tr><td><table><tr><th>x</th><th>y</th></tr><tr><td>1</td><td>1</td></tr><tr><td>2</td><td>8</td></tr><tr><td>3</td><td>27</td></tr><tr><td>4</td><td>64</td></tr><tr><td>5</td><td>125</td></tr><tr><td>6</td><td>216</td></tr><tr><td>7</td><td>?</td></tr></table></td><td>○</td><td>○</td></tr>
<tr><td>$y=-4x+12$</td><td>○</td><td>○</td></tr>
</table>

Date of Completion:______________ Score:______________

CHAPTER 2 → Lesson 4: Comparing Functions

1. A set of instructions says to subtract 5 from a number n and then double that result, calling the final result p. Which function rule represents this set of instructions?

Ⓐ $p = 2(n - 5)$
Ⓑ $p = 2n - 5$
Ⓒ $n = 2(p - 5)$
Ⓓ $n = 2p - 5$

2. Which of the following linear functions is represented by the (x, y) pairs shown in the table below?

x	y
-3	-1
1	7
4	13

Ⓐ $y = x + 2$
Ⓑ $y = 2x + 5$
Ⓒ $y = 3x + 1$
Ⓓ $y = 4x + 3$

3.

x	y
0	3
1	5
2	7

Three (x, y) pairs of a linear function are shown in the table above. Which of the following functions has the same slope as the function shown in the table?

Ⓐ $y = 3x + 2$
Ⓑ $y = 2x - 4$
Ⓒ $y = x + 3$
Ⓓ $y = x - 1$

4. The graph of linear function A passes through the point (5, 6). The graph of linear function B passes through the point (6, 7). The two graphs intersect at the point (2, 5). Which of the following statements is true?

Ⓐ Function A has the greater slope.
Ⓑ Both functions have the same slope.
Ⓒ Function B has the greater slope.
Ⓓ No relationship between the slopes of the lines can be determined from this information.

5. If line M includes the points (-1, 4) and (7, 9) and line N includes the points (5, 2) and (-3, -3), which of the following describes the relationship between M and N?

Ⓐ They are perpendicular.
Ⓑ They intersect but are not perpendicular.
Ⓒ They are parallel.
Ⓓ Not enough information is provided.

6. If line R includes the points (-2, -2) and (6, 4) and line S includes the points(0,4) and (3,0) which of the following describes the relationship between R and S?

Ⓐ They are perpendicular.
Ⓑ They intersect but are not perpendicular.
Ⓒ They are parallel.
Ⓓ Not enough information is provided.

7.

x	y
-4	13
2	1
6	-7

Three points of a linear function are shown in the table above. What is the y-intercept of this function?

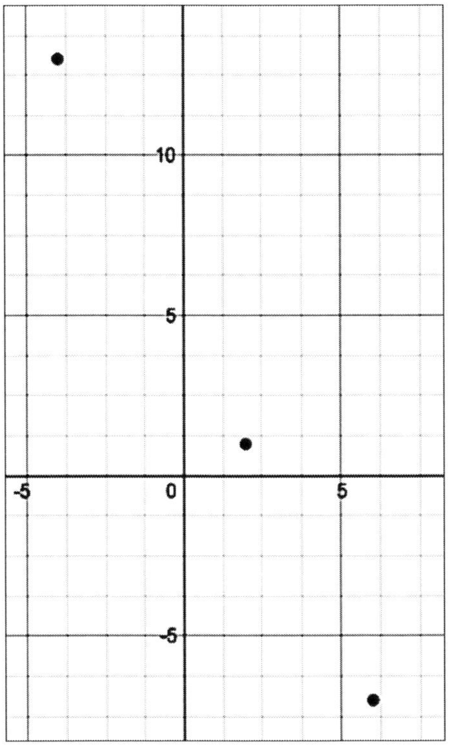

Ⓐ (0, 5)
Ⓑ (0, 6)
Ⓒ (0, 7)
Ⓓ (0, 8)

8. Comparing the two linear functions, y = 2x + 7 and y = 7x + 2, we find that

Ⓐ y = 2x + 7 has the steeper slope.
Ⓑ The graphs of the functions will be parallel lines.
Ⓒ The graphs of the functions will intersect at the point (1, 9).
Ⓓ The graphs of the functions will be perpendicular lines.

9. **Which of the following is true of the graphs of the lines $y = 3x + 5$ and $y = -\frac{x}{3} + 6$?**

Ⓐ **The graphs will be two parallel straight lines.**
Ⓑ **The graphs will be two perpendicular straight lines.**
Ⓒ **Both functions will produce the same line.**
Ⓓ **There is not enough information to compare.**

10. **If line A, denoted by $y = x + 9$, and line B, denoted by $y = 5x - 3$ are graphed, which of the following statements is correct?**

Ⓐ **Line B has a steeper slope.**
Ⓑ **Both lines have the same slope.**
Ⓒ **Line A has the steeper slope.**
Ⓓ **There is not enough information to answer the question.**

11. **A function may be described in 3 ways : (1) Set of ordered pairs (x,y) is given or (2) An equation showing how y depends on x or (3) a table showing how y varies with x. In the problem given below, take x to be the time and y to be the distance travelled.**

Select the ones that have the correct conclusion. Note that more than one option may be correct.

Ⓐ **CJ' s Speed:**

X	Y
0	0
2	13
4	26

Holly's Speed: $y=6x$ Conclusion: Holly is faster than CJ.

Ⓑ **Hank's Speed:** $y= \frac{7}{2}x$ **Smith's Speed:** {(0,0),(4,5),(8,10)} Conclusion: Hank is faster than Smith.

Ⓒ **Tyler's Speed:**

X	Y
0	0
4	6
8	12

Omar's Speed: $y= \frac{3}{2}x$ Conclusion: Tyler and Omar run the same speed.

12. Put the correct inequality or equality sign between the rates of change of the two functions given below.

Instruction : Take f to be the first function {(0, 0), (3, 180), (6, 360), (9, 540)} and g to be the second function y = 100x

{(0, 0), (3, 180), (6, 360), (9, 540)} [ ] y = 100x

13. Match each function to whether the rate of change is positive or negative

	Positive	Negative
$y=3x+4$	○	○
$y=\frac{1}{2}x-6$	○	○
$y=-7x+2$	○	○
$y=\frac{2}{3}x-1$	○	○

Date of Completion:_______________ Score:_______________

CHAPTER 2 → Lesson 5: Analyzing Functions

1. Complete the following:
The cost per copy is a function of the number of copies of any one title purchased.
This implies that _______________

Ⓐ the cost per copy of any one title is always a constant.
Ⓑ the cost per copy of any one title will change based on the number of copies purchased.
Ⓒ the cost per copy of any one title is not related to the number of copies purchased.
Ⓓ None of the above.

2. If a student's math grade is a positive function of the number of hours he spends preparing for a test, which of the following is correct?

Ⓐ The more he studies, the lower his grade.
Ⓑ The more he studies, the higher his grade.
Ⓒ There is no relation between how much he studies and his grade.
Ⓓ The faster he finishes his work, the higher his grade will be.

3. Mandy took a math quiz and received an initial score of i. She retook the quiz several times and, with each attempt, doubled her previous score.
After a TOTAL of four attempts, her final score was _______.

Ⓐ 2i
Ⓑ 3i
Ⓒ 2^3i
Ⓓ None of the above.

4. The graph of a linear function lies in the first and third quadrants. Which of the following must be true?

Ⓐ It also lies in the second quadrant.
Ⓑ It also lies in the fourth quadrant.
Ⓒ It is an increasing function.
Ⓓ It is a decreasing function.

5. The graph of an increasing linear function crosses the vertical axis at (0, -1). Which of the following CANNOT be true?

Ⓐ It also lies in the second quadrant.
Ⓑ It also lies in the fourth quadrant.
Ⓒ It is an increasing function.
Ⓓ y intercept is -1.

6. A linear equation is plotted on the coordinate plane, and its graph is perpendicular to the x-axis. Which of the following best describes the slope of this line?

Ⓐ Zero
Ⓑ Undefined
Ⓒ Negative
Ⓓ Positive

7. The graph of a decreasing linear function crosses the vertical axis at (0, 3). Which of the following CANNOT be true?

Ⓐ The graph lies in the first quadrant.
Ⓑ The graph lies in the second quadrant.
Ⓒ The graph lies in the third quadrant.
Ⓓ The graph lies in the fourth quadrant.

8. Which of the following best describes the x & y coordinates of any point in the first quadrant?

Ⓐ Both are positive
Ⓑ Both are negative
Ⓒ One is positive and one is negative
Ⓓ None of these

9. In the coordinate plane, in which quadrant is the ordered pair, (-3, -6) located?

Ⓐ I
Ⓑ II
Ⓒ III
Ⓓ IV

10. In the coordinate plane, in which quadrant is the ordered pair, (1, 8) located?

Ⓐ I
Ⓑ II
Ⓒ III
Ⓓ IV

11. In the coordinate plane, in which quadrant is the ordered pair, (6, -9) located?

Ⓐ I
Ⓑ II
Ⓒ III
Ⓓ IV

12. A linear equation is plotted on the coordinate plane, and its graph is parallel to the x-axis. Which of the following best describes the slope of this line?

Ⓐ Zero
Ⓑ Undefined
Ⓒ Negative
Ⓓ Positive

13. Which of the following best describes the x & y coordinates of any point in the third quadrant?

Ⓐ Both are positive
Ⓑ Both are negative
Ⓒ One is positive and one is negative
Ⓓ None of these

14. In a coordinate plane the graphs of two functions are perpendicular. Which of the following best describes the relationship of their slopes?

Ⓐ Their slopes are the same.
Ⓑ Their slopes are both zero.
Ⓒ Their slopes are both one.
Ⓓ One slope is the negative reciprocal of the other.

15. In a coordinate plane the graphs of two linear functions are parallel lines. Which of the following best describes the relationship of their slopes?

Ⓐ Their slopes are the same.
Ⓑ Their slopes are opposites.
Ⓒ The slope of one line is the negative reciprocal of the other.
Ⓓ There is no known relationship between their slopes.

16. Observe the graph given.
Match each segment to whether it is increasing or decreasing as per the graph.

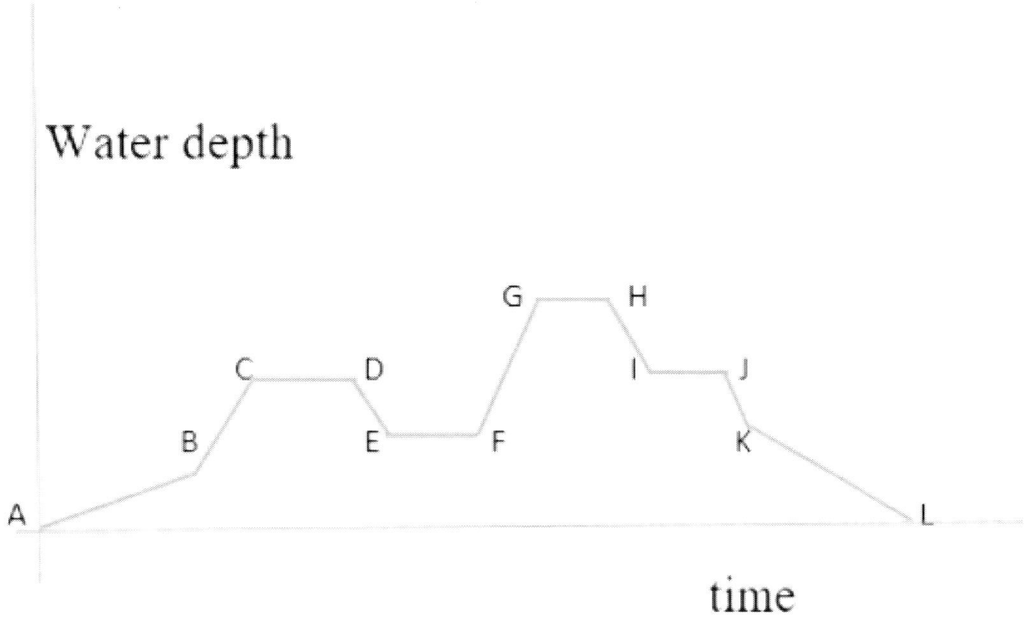

	INCREASING	DECREASING	CONSTANT
A to B	◯	◯	◯
B to C	◯	◯	◯
G to H	◯	◯	◯
I to J	◯	◯	◯
J to K	◯	◯	◯
K to L	◯	◯	◯
C to D	◯	◯	◯
E to F	◯	◯	◯
D to E	◯	◯	◯
F to G	◯	◯	◯
H to I	◯	◯	◯

17. In which interval is this graph increasing?

Write your answer in the box given below

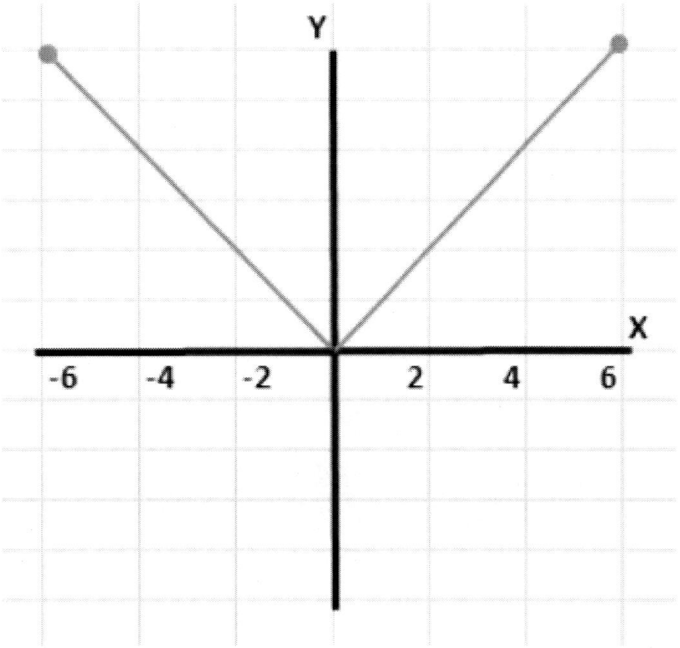

18. Match each equation with whether or not the function is increasing, decreasing, or constant.

	INCREASING	DECREASING	CONSTANT
$y=-2x+5$	○	○	○
$y=6$	○	○	○
$y=-\frac{1}{2}x$	○	○	○
$y=5x-9$	○	○	○

CHAPTER 2 → Lesson 6: Analyzing Linear Scatter Plots

1.

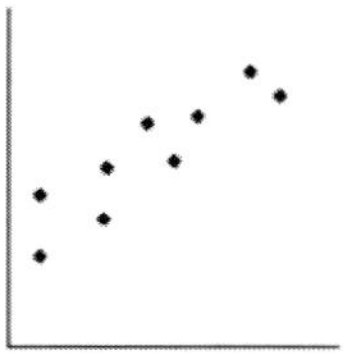

Which of the following scatter plots below demonstrates the same type of data correlation as the one shown above?

Ⓐ 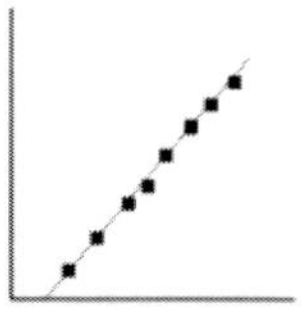

Ⓑ 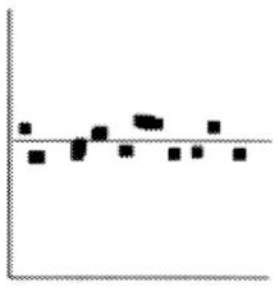

Ⓒ 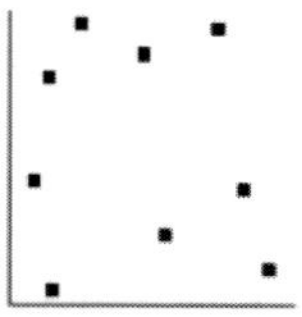

Ⓓ 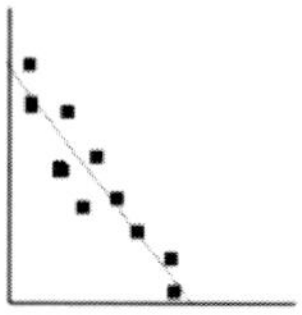

2.

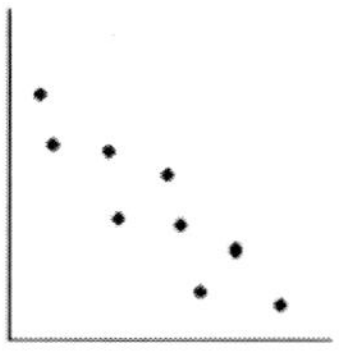

Which of the following lines most accurately models the points in this scatter plot?

Ⓐ 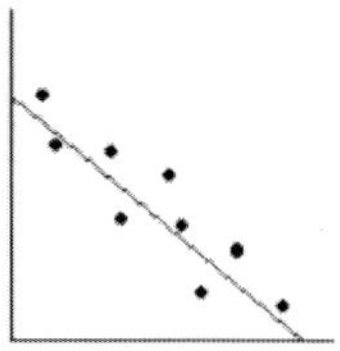

Ⓑ 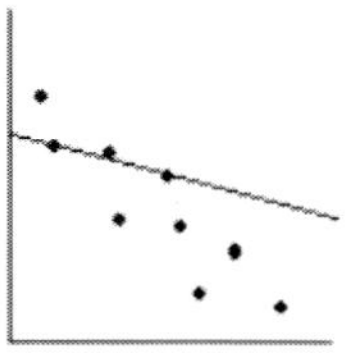

Ⓒ 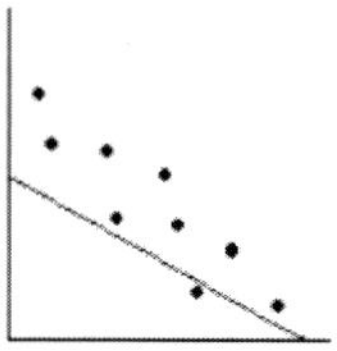

Ⓓ 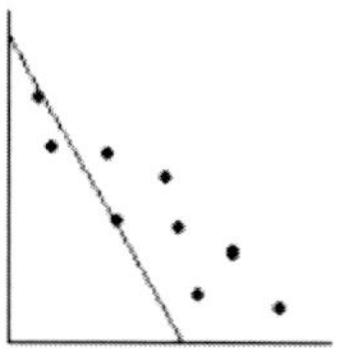

3. 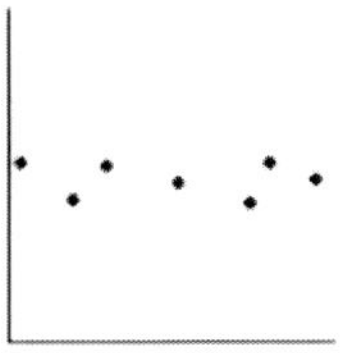

Which of the following lines most accurately models the points in this scatter plot?

Ⓐ 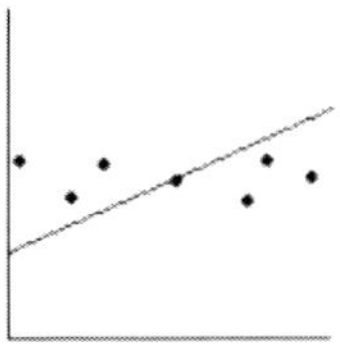

Ⓑ 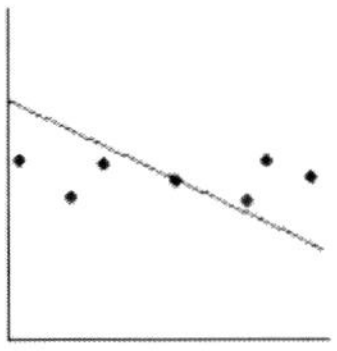

Ⓒ 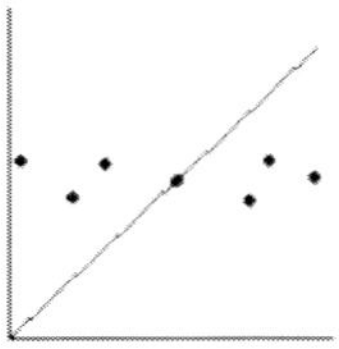

Ⓓ 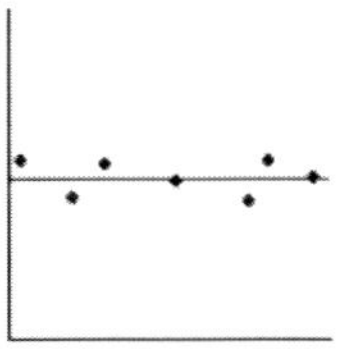

4. 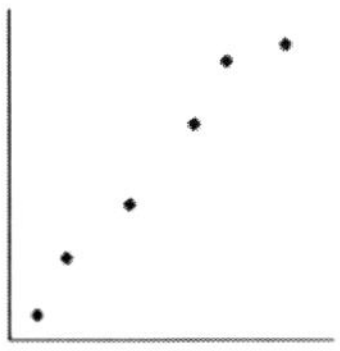

Which of the following lines most accurately models the points in this scatter plot?

Ⓐ 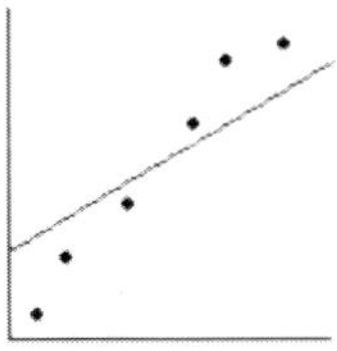

Ⓑ 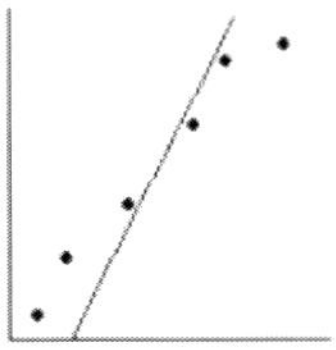

Ⓒ 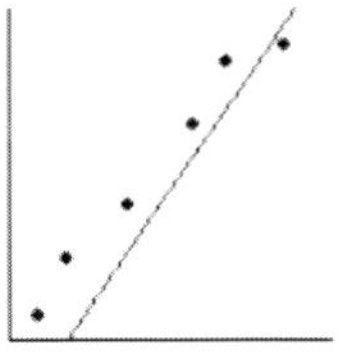

Ⓓ 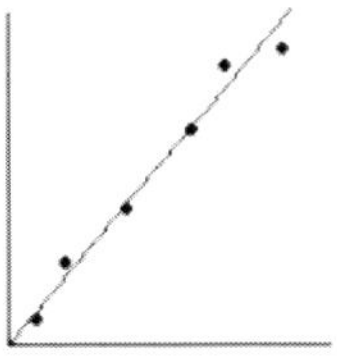

5. **The four scatter plots below have all been modeled by the same line. Which scatter plot has the strongest association?**

Ⓐ 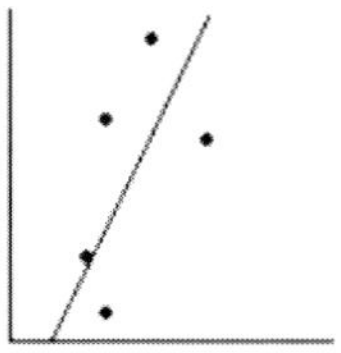

Ⓒ 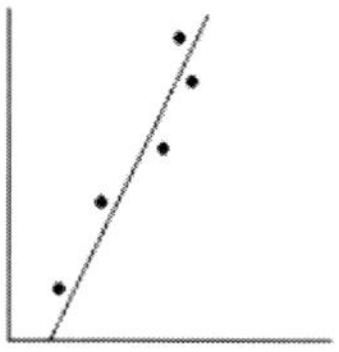

Ⓑ 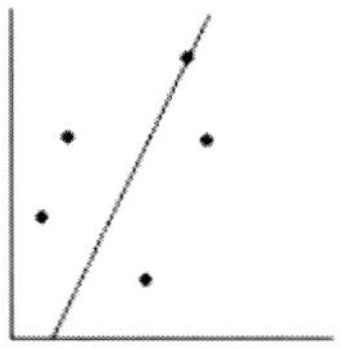

Ⓓ 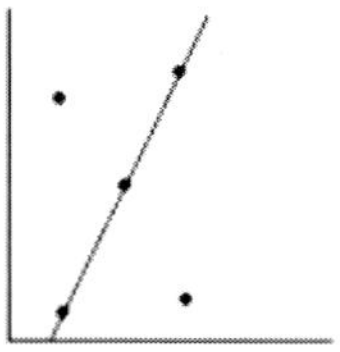

6. **The four scatter plots shown below have four points in common, and each scatter plot has a different fifth point. Which scatter plot's fifth point is an outlier?**

Ⓐ 

Ⓒ 

Ⓑ 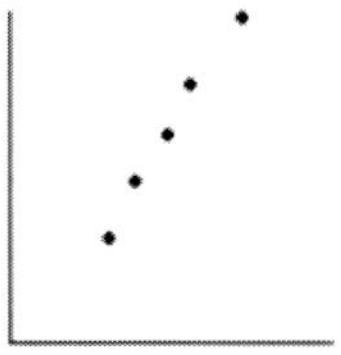

Ⓓ 

7. **The four scatter plots shown below have four points in common, and each scatter plot has a different fifth point. Which scatter plot's fifth point is NOT an outlier?**

Ⓐ 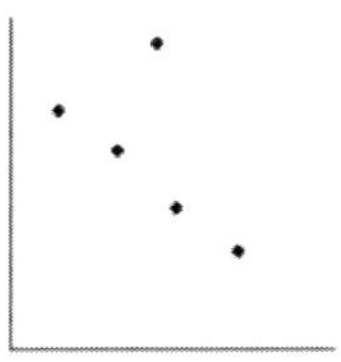

Ⓒ 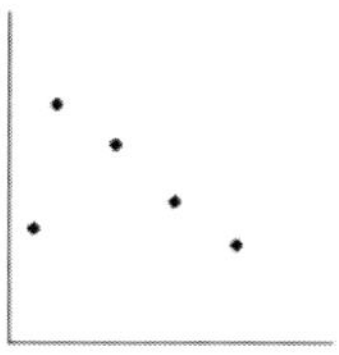

Ⓑ 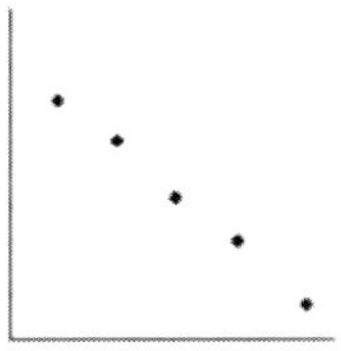

Ⓓ 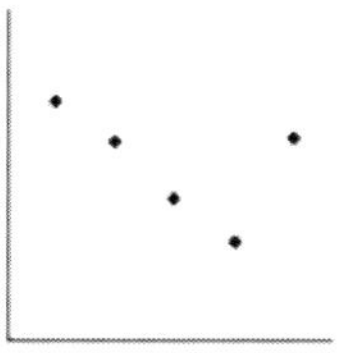

8. **The figure below shows a scatter plot relating the length of a bean plant, in centimeters, to the number of days since it was planted. The slope of the associated line is 2. Which of the following correctly interprets the slope?**

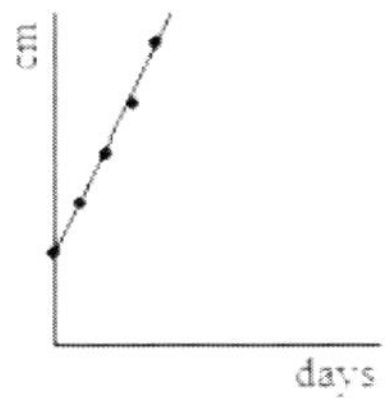

Ⓐ **The bean plant grows approximately 1 cm every 2 days.**
Ⓑ **The bean plant grows approximately 2 cm each day.**
Ⓒ **The bean plant was 2 cm long when it was planted.**
Ⓓ **The bean plant approximately doubles in length each day.**

9. The figure below shows a scatter plot relating the cost of a ride in a taxicab, in dollars, to the number of miles traveled. The slope of the associated line is 0.5. Which of the following correctly interprets the slope?

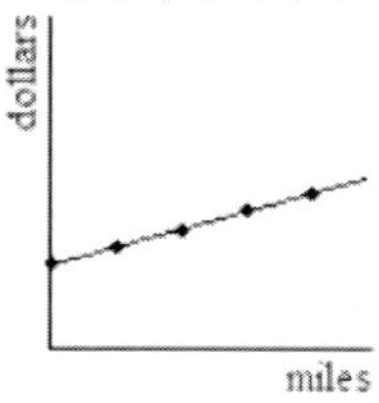

Ⓐ For each additional mile traveled, the cost of the ride increases by 50 cents.
Ⓑ For each additional half of a mile traveled, the cost of the ride increases by 1 dollar.
Ⓒ The initial cost of the ride, before the taxi has traveled any distance, is 50 cents.
Ⓓ The first half of a mile does not cost anything.

10. The figure below shows a scatter plot relating the temperature in a school's parking lot, in degrees Fahrenheit, to the number of hours past noon. The slope of the associated line is -3. Which of the following correctly interprets the slope?

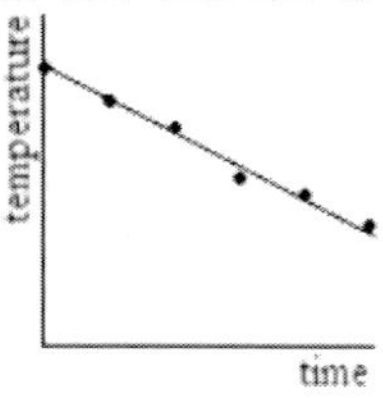

Ⓐ The temperature at noon was -3 degrees Fahrenheit.
Ⓑ The temperature decreased until it reached -3 degrees Fahrenheit.
Ⓒ The temperature decreased an average of 1 degree Fahrenheit every 3 hours.
Ⓓ The temperature decreased an average of 3 degrees Fahrenheit per hour.

11. Match the correct vocab term with the correct definition.

	Linear	Negative Association	Line of Best Fit	Prediction Equation
A line on a graph showing the general direction that a group of points seem to be heading	○	○	○	○
A graph that is represented by a straight line	○	○	○	○
The equation of a line that can predict outcomes using given data	○	○	○	○
A correlation of points that is linear with a negative slope	○	○	○	○

12. Write the equation of the best fit line for this scatter plot using the 2 ordered pairs given.

Write your answer in the box given below.

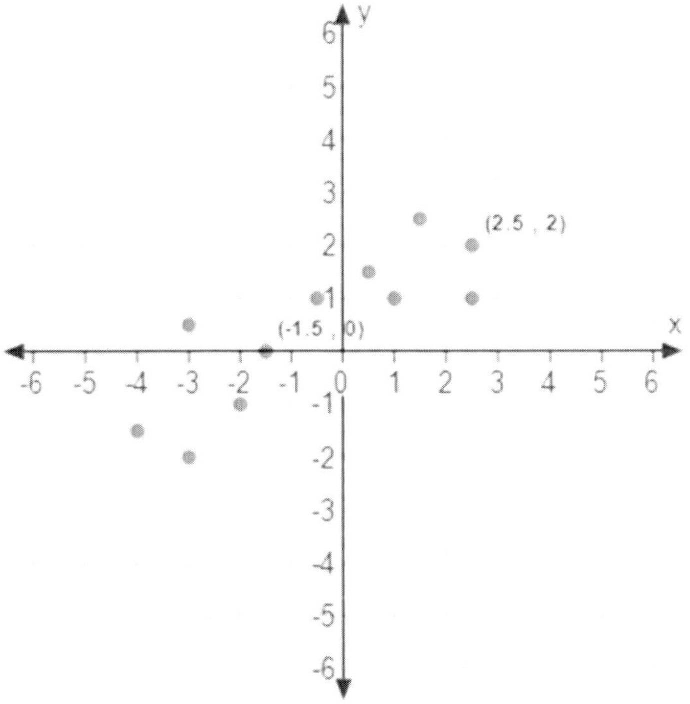

CHAPTER 2 → Lesson 7: Solving Linear Equations

1. Which two consecutive odd integers have a sum of 44?

Ⓐ 21 and 23
Ⓑ 19 and 21
Ⓒ 23 and 25
Ⓓ 17 and 19

2. During each of the first three quarters of the school year, Melissa earned a grade point average of 2.1, 2.9, and 3.1. What does her 4th quarter grade point average need to be in order to raise her grade to a 3.0 cumulative grade point average?

Ⓐ 3.9
Ⓑ 4.2
Ⓒ 2.6
Ⓓ 3.5

3. Martha is on a trip of 1,924 miles. She has already traveled 490 miles. She has 3 days left on her trip. How many miles does she need to travel each day to complete her trip?

Ⓐ 450 miles/day
Ⓑ 464 miles/day
Ⓒ 478 miles/day
Ⓓ 492 miles/day

4. Find the solution to the following equation: $3x + 5 = 29$

Ⓐ $x = 24$
Ⓑ $x = 11$
Ⓒ $x = 8$
Ⓓ $x = 6$

5. Find the solution to the following equation:
$7 - 2x = 13 - 2x$

Ⓐ $x = -10$
Ⓑ $x = -3$
Ⓒ $x = 3$
Ⓓ There is no solution.

6. Find the solution to the following equation: 6x + 1 = 4x - 3

Ⓐ x = -1
Ⓑ x = -2
Ⓒ x = - 0.5
Ⓓ There is no solution.

7. Find the solution to the following equation:
2x + 6 + 1 = 7 + 2x

Ⓐ x = -3
Ⓑ x = 3
Ⓒ x = 7
Ⓓ All real numbers are solutions.

8. Find the solution to the following equation: 8x-1=8x

Ⓐ x = 7
Ⓑ x = -8
Ⓒ x = 8
Ⓓ There is no solution.

9. Which of the answers is the correct solution to the following equation?
2x + 5x - 9 = 8x - x - 3 - 6

Ⓐ x = 3
Ⓑ x = 7
Ⓒ x = 9
Ⓓ All real values for x are correct solutions.

10. Solve the following linear equation for y.
-y + 7y -54 = 0

Ⓐ y = 0
Ⓑ y = 1
Ⓒ y = 6
Ⓓ y = 9

11. Solve each equation for the variable. Select the ones whose values of the variables are the same.

Note that more than one option may be correct. Select all the correct answers.

Ⓐ -5m = 25
Ⓑ -10c = -80
Ⓒ -7 + g = -12
Ⓓ 12m + 20 = -40

12. Solve for x: 5x + 20 = -20.

x = ?

Write your answer in the box given below.

13. Fill in the missing number to make this equation true, if x = -3.

7+3x=5x+______

CHAPTER 2 → Lesson 8: Solve Linear Equations with Rational Numbers

1. Solve the following linear equation: $\frac{7}{14} = n + \frac{7}{14}n$

Ⓐ $n = 1\frac{1}{2}$

Ⓑ $n = 3$

Ⓒ $n = \frac{1}{3}$

Ⓓ $n = 1$

2. Find the solution to the following equation: $2(2x - 7) = 14$

Ⓐ $x = 14$
Ⓑ $x = 7$
Ⓒ $x = 1$
Ⓓ $x = 0$

3. Solve the following equation for x.
$6x - (2x + 5) = 11$

Ⓐ $x = -3$
Ⓑ $x = -4$
Ⓒ $x = 3$
Ⓓ $x = 4$

4. $4x + 2(x - 3) = 0$

Ⓐ $x = 0$
Ⓑ $x = 1$
Ⓒ $x = 2$
Ⓓ All real values for x are correct solutions.

5. Solve the following equation for y.
$3y - 7(y + 5) = y - 35$

Ⓐ $y = 0$
Ⓑ $y = 1$
Ⓒ $y = 2$
Ⓓ All real values for y are correct solutions.

6. Solve the following linear equation: $2(x-5) = \frac{1}{2}(6x+4)$

Ⓐ $x= -12$
Ⓑ $x= -9$
Ⓒ $x= -4$
Ⓓ There is no solution.

7. Solve the following linear equation for x.

$3x + 2 + x = \frac{1}{3}(12x + 6)$

Ⓐ $x= -4$
Ⓑ $x= 2$
Ⓒ There is no solution.
Ⓓ All real values for x are correct solutions.

8. $\frac{1}{2}x + \frac{2}{3}x + 5 = \frac{5}{2}x + 6$

Ⓐ $x = \frac{33}{4}$

Ⓑ $x = \frac{1}{2}$

Ⓒ $x =- \frac{3}{4}$

Ⓓ $x =- \frac{6}{5}$

9. Which of the following could be a correct procedure for solving the equation below?
$2(2x+3) = 3(2x+5)$

Ⓐ $4x+5 = 6x+5$
$-2x+5 = 5$
$-2x = 0$
$x = 0$

Ⓑ $4x+6 = 6x+15$
$-2x+6 = 15$
$-2x = 9$
$x = -\frac{9}{2}$

Ⓒ $2(5x) = 6x+15$
$10x = 6x+15$
$4x = 15$
$x = \frac{15}{4}$

Ⓓ $4x+6 = 6x+15$
$-2x+6 = 15$
$-2x = 9$
$x = \frac{2}{9}$

10. Solve the following linear equation:
$0.64x - 0.15x + 0.08 = 0.09x$

Ⓐ $x = -5$
Ⓑ $x = -0.2$
Ⓒ $x = 5.125$
Ⓓ There is no solution.

11. Which of the following statements are true? Select all that apply.

Ⓐ $w - \frac{2}{5} = \frac{8}{5}$ so $w=2$

Ⓑ $\frac{-5}{8} y=15$ so $y=24$

Ⓒ $0.4x-1.2 =0.15x+0.8$ so $x=8$

Ⓓ $\frac{x}{6} =-5$ so $x=30$

12. Select the answer choice with the correct order of how you would solve the equation.

$\frac{3}{4}+\frac{1}{2}(m+\frac{1}{4})=\frac{19}{16}$

A --> $m=\frac{5}{8}$ B --> $\frac{1}{2}(m+\frac{1}{4})=\frac{7}{16}$ C --> $(m+\frac{1}{4})=\frac{7}{8}$

Ⓐ A B C
Ⓑ B C A
Ⓒ C A B

13. Solve the following linear equation:

$\frac{8}{16}= n+\frac{8}{16} n$

End of Algebraic Reasoning & Algebra

ANSWER KEY AND DETAILED EXPLANATION

Chapter 2
Algebraic Reasoning & Algebra

Lesson 1: Functions

Question No.	Answer	Detailed Explanation
1	D	In order to be classified as a function each x must map to one and only one value of y. In the case of the set of ordered pairs {(2, 3), (2, 7), (8, 6)}, the x value of 2 has two different values for y, 3 and 7. Therefore, this set does not classify as a function.
2	B	In the case of <table><tr><th>x</th><th>y</th></tr><tr><td>1</td><td>7</td></tr><tr><td>3</td><td>8</td></tr><tr><td>4</td><td>7</td></tr></table>for each x there is one and only one value for y. Therefore, this table represents a function.
3	A	If y is a function of x, a particular x value CANNOT be associated with two or more values of y.
4	A	In the first graph, for a certain value of x, there are two distinctly different values for y. This is NOT a function.
5	C	In the third graph for certain values of x, there are more than one value for y. This is NOT a function
6	D	{(0, 0), (0, 2), (0, 4)} does not qualify as a function because for x = 0, there is more than one value for y.
7	D	A line drawn through the graph perpendicular to the x-axis will cross the graph one and only one time because, for each value of x, there will be one and only one value of y.

Question No.	Answer	Detailed Explanation
8	D	In a function, for each value of x, there must be one and only one value for y. We already have (2,1) so cannot have another ordered pair where x=2.
9	D	There is already an assigned y value for x=6, x=5, and x=4, but not for x = 3.
10	D	Only this set, [(-7,10), (7,10), (8,9), (9,10)], represents a function because for each value for x, there exists one and only one value for y.
11	A, C	If a set of ordered pairs is a function then it will have each x-value going to only one y-value. If an x-value is paired with more than one y-value, then it is not a function.
12	y=x-3	The rule for this function is y=x-3. To get this you find the slope first. In this case the slope is 1. Then you find out when the y is zero. In this case it is when x is 3. Thus combining them to form the rule.
13		It is a function if each x-value is paired with only one y-value. If y depends linearly on x, in the form, y = mx + b, then also it is a function. Because, in such cases too, for each value of x, there is only one value of y.

Lesson 2: Linear Function Models

Question No.	Answer	Detailed Explanation
1	D	There is not enough information to say any of the choices MUST be true although it may represent a portion of the graph of a constant function.
2	A	$y = mx + 2$ $0 = 4m + 2$ $-2 = 4m$ $-\frac{2}{4} = m$ $-\frac{1}{2} = m$ Therefore, m is negative is true.
3	A	$y = 2x + b$ $0 = 2(-5) + b$ $0 = -10 + b$ $10 = b$ Therefore, b is positive is true.
4	D	All that we know is that the y-intercept is (0,0); i.e. b=0.
5	C	$m = \frac{5 - 5}{9 - 2} = \frac{0}{7} = 0$ $y = mx + b$ $5 = 0 + b$ $5 = b$
6	C	$m = \frac{7 - 5}{2 - 3} = \frac{2}{-1} = -2$
7	C	It is linear and the y value is 1 in the case of the two given points. Therefore, it is a constant function.
8	A	The slopes range between $\frac{1}{2}$ and 2; so 2 is the greatest slope. The first option is the correct answer.
9	B	Height of the bench is 18 in. The height increases by 3 in. for each block added; so $h = 3b + 18$

Question No.	Answer	Detailed Explanation
10	C	He starts out $10 in debt so b = -10. At any point, he has earned 5w where w is the number of weeks. Therefore, m = 5w -10.
11	A	A guest must pay $5 admission before any rides; so b=5. He pays $2 per ride so the total paid in rides is 2r. Now add these two amounts and p = 2r + 5
12	D	The slopes range between $\frac{1}{3}$ and $\frac{4}{3}$; so $\frac{1}{3}$ is the smallest slope. The last choice is the correct answer.
13	B	y = mx + b $m = \frac{4 - 2}{9 - 6} = \frac{2}{3}$ Substitute (6,2). $2 = 6(\frac{2}{3}) + b$ 2 = 4 + b 2-4 =b -2 = b Therefore, y-intercept is (0, -2)
14	D	The point (0,-5) tells us that b = -5. $m = \frac{15 - 0}{4 - 1} = \frac{15}{3} = 5$ Substitute into y = mx + b. f(x) = 5x - 5
15	B	Since you pay $20 for the first CD, b = 20. If you purchase x CDs and you already paid $20 for the first one, the remainder would be (x-1) CDs for which you pay $10 each. Therefore, f(x) = 10(x-1) + 20.
16	3	To find the missing information, we are looking for the rate of change. You will fill this into the slope formula. $\frac{220 - 178}{90 - 76}$ =3. Thus the equation will be y=3x-50.

<table>
<tr><th>Question No.</th><th>Answer</th><th>Detailed Explanation</th></tr>
<tr><td>17</td><td></td><td>
<table>
<tr><th></th><th>y=−4x−3</th><th>y=7x-5</th><th>y=-x+2</th></tr>
<tr><td>(1, 2) slope=7</td><td>○</td><td>●</td><td>○</td></tr>
<tr><td>(-2, 5) slope=-4</td><td>●</td><td>○</td><td>○</td></tr>
<tr><td>(3, -1) slope=-1</td><td>○</td><td>○</td><td>●</td></tr>
</table>

Using the slope and an ordered pair you will have to solve for the y-intercept. To do this you will substitute the ordered pair in for x and y into an equation in the format y=mx+b. Once you have this substituted in you will solve for b. This will then give you the equation.

Alternate Explanation : Slope of the line y = mx + b is m (coefficient of x). In this particular problem, since the slopes of the three lines are different, you can easily guess the equations without doing any calculation.</td></tr>
<tr><td>18</td><td>$19.99</td><td>In this case the slope or rate of change is $19.99.</td></tr>
<tr><td>19</td><td>[{"x":0,"y":11},
{"x":2,"y":7},
{"x":3,"y":5}]</td><td>Coordinates are represented as (x,y)

(0,11) means 0 on x axis and 11 on y axis

(2,7) means 2 on x axis and 7 on y axis.

(3,5) means 3 on x axis and 5 on y axis.

Plot the points.</td></tr>
</table>

Lesson 3: Linear Functions

Question No.	Answer	Detailed Explanation
1	D	The slope of the line through the given points is $\frac{1}{2}$. So slope (m) = $\frac{11 - 7}{k - 6} = \frac{1}{2}$ k-6 = 2(11-7) k-6 = 8 k = 14
2	C	In linear equations, none of the variables is of the second power or higher and they can be put into the form of y = mx + b.
3	B	In the function f(x) = x - 3, the coordinates of (3,0) satisfy the equation and it is written in the form of a linear equation.
4	A	The slope is a constant $\frac{3}{2}$ between any two points and slope is positive; so y increases as x increases. Therefore, it is increasing linearly.
5	D	Only the last statement, the function is nonlinear because it does not have the same slope between different airs of points, is true.
6	B	If it is a function, x has to change as y changes, but if that is true, the line cannot be constant for y unless it is in quadrants I and II, or III and IV, or it is the x-axis.
7	C	If the function is linear, the slope has to be the same between any 2 points. Using the first two points, the slope is $\frac{2}{3}$. Therefore, using the next two points, the slope must also be $\frac{2}{3}$; so $\frac{n - 3}{9 - 3} = \frac{2}{3}$; so 3n-9=18-6 3n=12+9=21 n=7

Question No.	Answer	Detailed Explanation
8	C	$\frac{9-3}{3-0}=\frac{n-9}{9-3}$ $\frac{6}{3}=\frac{n-9}{6}$ $\frac{2}{1}=\frac{n-9}{6}$ n-9=12 n=21
9	D	If the graph lies in quadrants I and II and is non-negative, it cannot reach quadrant IV, although it could also pass through quadrant III.
10	A	Since the slope is 2, y = 2x + 2 is the only function that could fit.
11	B, C	If the graph is a curved line, or the function has a power of x other than 1, then the function is not linear. Therefore, option (A) and (D) are not linear. Option (B) has a power of x equal to 1 only. Therefore, it is linear. Calculate the ratio of change in y to change in x. If the ratio is same for all points, then the graph is a straight line and the function is linear. Option (C) satisfies this condition. Therefore, it is linear.
12	Non Linear	This function will be non-linear because if we rearrange to solve for y, you will still have x to the second power. Thus making it non-linear.
13		If the function has a power of x other than 1, then the function is non-linear. Therefore, first function is non-linear. Second function is a constant function. It is a linear function. Calculate the ratio of change in y to change in x. If the ratio is same for all points, then the graph is a straight line and the function is linear. The table in the third option does not satisfy this condition. Therefore it is a non-liner function. Fourth function has a power of x equal to 1 only. Therefore, it is linear.

Lesson 4: Comparing Functions

Question No.	Answer	Detailed Explanation
1	A	Only p = 2(n – 5) meets the requirements to subtract 5 from n and then double the result to get p.
2	B	The slope is 2. The only function with a slope of 2 is y = 2x + 5.
3	B	According to the table, when x increases by 1, y increases by 2. Therefore the slope of the line (m) is 2. The only function offered with a slope of 2 is y = 2x - 4.
4	C	Function A has a slope of $\frac{1}{3}$ and Function B has a slope of $\frac{1}{2}$. Therefore, Function B has a greater slope. To verify, apply the formula: Slope = m =$(y_2-y_1)/(x_2-x_1)$ to each function.
5	C	The slope (m) of Line A = $\frac{9-4}{7-(-1)} = \frac{5}{8}$ The slope (m) of Line B = $\frac{-3-2}{-3-5} = \frac{-5}{-8} = \frac{5}{8}$ Since their slopes are equal, the lines are parallel.
6	A	The slope (m) of Line R = $\frac{4+2}{6+2} = \frac{6}{8} = \frac{3}{4}$ The slope (m) of Line S = $\frac{4-0}{0-3} = \frac{4}{-3}$ Since their slopes are negative reciprocals of each other, the lines are perpendicular.
7	A	y = mx + b $m = \frac{13-1}{-4-2} = \frac{12}{-6} = -2$ y = -2x + b Substitute (x,y) for one point and solve for b. Using (2,1), we get 1 = -2(2) + b 1 = -4 + b 1 + 4 = b 5 = b; so y-intercept is (0,5)
8	C	If you substitute the coordinates of the point (1,9) into each equation, you will find that both equations are satisfied. "The graphs of the functions will intersect at the point (1,9)" is the correct answer.

Question No.	Answer	Detailed Explanation
9	B	Since their slopes, 3 and $-\frac{1}{3}$, are negative reciprocals of each other, the lines will be perpendicular.
10	A	The slope of line A is 1 and that of line B is 5. Therefore, line B has a steeper slope.
11	B, C	These are showing different representations of a function. In comparing them you need to look at their slopes. Whichever one is higher represents the faster person. In the equations, the slope is found by looking at the number that is paired with the x. From the table or the ordered pairs you have to use the slope formula. (A) CJ's speed = (13-0) / (2-0) = 13/2. Holly's speed = 6. Therefore CJ's speed > Holly's speed. Conclusion is wrong. (B) Hank's speed = 7/2. Smith's speed = (5-0) / (4-0) = 5/4. Therefore Hank's speed > Smith's speed. Conclusion is correct. (C) Tyler's speed = (6-0) / (4-0) = 6/4 = 3/2. Omar's speed = 3/2. Therefore Tyler's speed = Omar's speed. Conclusion is correct.
12	Rate of change of function f < [Rate of change of function g]	For the function y = 100x, you know the rate of change is 100, so you need to find out the rate of change of the other one. To do this you will have to find the slope, $\frac{180-0}{3-0}$ which is equal to 60. So in this case we are comparing 60 and 100. Since 60 is smaller than 100 the inequality symbol should be < (less than) for this situation.
13		To determine whether the rate of change is positive or negative you have to look at the slope. The slope is the number paired with the x. If it is a positive number, then the rate of change is positive. If the number is negative, then the rate of change is negative.

Lesson 5: Analyzing Functions

Question No.	Answer	Detailed Explanation
1	B	The cost per copy is a function of the number of copies purchased means that the cost per copy changes as the number of copies changes.
2	B	In a positive function, as one variable increases so does the other.
3	C	Her first score was *i*. She doubled the previous score three times or 2^3i.
4	C	The line could also lie in the second or fourth quadrant, but it definitely has a positive slope and is, consequently, an increasing function.
5	A	The graph cannot lie in the second quadrant.
6	B	m = change in y / change in x. In this case change in x is 0. Division by 0 is undefined.
7	C	The graph cannot lie in the third quadrant.
8	A	In the first quadrant, both x and y are positive.
9	C	If both coordinates are negative, the point lies in Quadrant III.
10	A	If both coordinates are positive, the point lies in Quadrant I.
11	D	If the x coordinate is positive, the point lies on the right side of the y-axis and if the y coordinate is negative, the point lies below the x-axis. This positions the point in Quadrant IV.
12	A	m = change in y / change in x. If line is parallel to x-axis, change in y is 0. If change in y is 0, m = 0 ÷ n = 0, where n is not zero, but is any other real number resulting from the change in x.
13	B	In the third quadrant, both coordinates are negative.
14	D	The slopes of perpendicular lines are negative reciprocals of each other.
15	A	The slopes of parallel lines are the same.

<table>
<tr><th>Question No.</th><th>Answer</th><th>Detailed Explanation</th></tr>
<tr><td>16</td><td></td><td>
<table>
<tr><th></th><th>INCREASING</th><th>DECREASING</th><th>CONSTANT</th></tr>
<tr><td>A to B</td><td>•</td><td></td><td></td></tr>
<tr><td>B to C</td><td>•</td><td></td><td></td></tr>
<tr><td>G to H</td><td></td><td></td><td>•</td></tr>
<tr><td>I to J</td><td></td><td></td><td>•</td></tr>
<tr><td>J to K</td><td></td><td>•</td><td></td></tr>
<tr><td>K to L</td><td></td><td>•</td><td></td></tr>
<tr><td>C to D</td><td></td><td></td><td>•</td></tr>
<tr><td>E to F</td><td></td><td></td><td>•</td></tr>
<tr><td>D to E</td><td></td><td>•</td><td></td></tr>
<tr><td>F to G</td><td>•</td><td></td><td></td></tr>
<tr><td>H to I</td><td></td><td>•</td><td></td></tr>
</table>
If a graph is increasing it will be going from a lower starting point to a higher ending point. If the segment of the graph is decreasing, then its starting point will be higher than the ending point. If the segment of the graph is constant, then the starting point and ending point will be at the same level.</td></tr>
<tr><td>17</td><td>$0 < X < 6$</td><td>Looking at this graph you can see that the graph is increasing from 0 to 6. This means that the starting point of the graph is lower than the ending point. So you would write the interval of $0 < x < 6$.</td></tr>
<tr><td>18</td><td></td><td>
<table>
<tr><th></th><th>INCREASING</th><th>DECREASING</th><th>CONSTANT</th></tr>
<tr><td>$y=-2x+5$</td><td></td><td>•</td><td></td></tr>
<tr><td>$y=6$</td><td></td><td></td><td>•</td></tr>
<tr><td>$y=-\frac{1}{2}x$</td><td></td><td>•</td><td></td></tr>
<tr><td>$y=5x-9$</td><td>•</td><td></td><td></td></tr>
</table>
When looking at just equations, you will have to look at the slope of the line to determine whether it is increasing, decreasing, or constant. The slope is the number that is paired with the x-value. If the slope is negative, then it is decreasing. If the slope is positive , the function is increasing. If there is no slope, then the line is horizontal, making the function a constant.</td></tr>
</table>

Lesson 6: Analyzing Linear Scatter Plots

Question No.	Answer	Detailed Explanation
1	A	Visual inspection of the points and lines shows that the first scatter plot is the best choice because it shows a positive association, just like the original scatter plot does. As x increases, y increases.
2	A	The line in choice one passes through the center of the scatter plot with approximately the same number of stray points on each side.
3	D	The line in choice 4 is clearly a best fit line.
4	D	The line in choice 4 passes through nearly every point in the scatter plot.
5	C	The line in choice 3 has the strongest association with the points in the scatter plot.
6	C	The fifth point in choice 3 is an outlier because it is positioned farther away from the line of best fit than in the other scatter plots.
7	B	Choices 1, 3, and 4 have outlier points. Choice 2 does not.
8	B	We are told that the slope is 2. The slope is always the change on the vertical axis divided by the change on the horizontal axis; so a slope of 2 would be interpreted as 2/1 which represents 2 units on the vertical axis for every unit on the horizontal axis. Looking at the graph we see that the vertical axis is cm and the horizontal axis is days. Therefore a slope of 2=2/1 would represent 2 cm growth for each day.
9	A	Slope represents the (change on the vertical axis) / (the change on the horizontal axis). The vertical axis represents dollars and the horizontal axis represents miles; so a slope of 0.5 would be interpreted as 0.5 dollars per mile or 0.5 dollars/mile which is 50 cents for each mile traveled.
10	D	We are told that the graph relates the temperature in a school's parking lot, in degrees Fahrenheit, to the number of hours past noon. A slope of -3/1 = a temperature decrease of 3°F/hour.

Question No.	Answer	Detailed Explanation
11		<table><tr><th></th><th>Linear</th><th>Negative Association</th><th>Line of Best Fit</th><th>Prediction Equation</th></tr><tr><td>A line on a graph showing the general direction that a group of points seem to be heading</td><td></td><td></td><td>●</td><td></td></tr><tr><td>A graph that is represented by a straight line</td><td>●</td><td></td><td></td><td></td></tr><tr><td>The equation of a line that can predict outcomes using given data</td><td></td><td></td><td></td><td>●</td></tr><tr><td>A correlation of points that is linear with a negative slope</td><td></td><td>●</td><td></td><td></td></tr></table> The line that is on the graph that doesn't necessarily go through every point but represents the general trend of the graph would be the line of best fit. If it is a straight line, then it is a linear graph. The equation of the best fit line that helps you predict outcomes are the prediction equation. Lastly, if the trend of the scatter plot has a negative slope then it has a negative association.
12	y=(1/2) x+3/4	Using the 2 points they gave us we first need to find the slope. Now that we know the slope, $\frac{2-0}{2.5-(-1.5)} = \frac{1}{2}$ we can fill in an ordered pair and solve for the y-intercept. $2 = \frac{1}{2}(2.5)+b$ When you solve for b you get 3/4. Put those numbers together and you get the equation y = (1/2)x + 3/4.

Lesson 7: Solving Linear Equations

Question No.	Answer	Detailed Explanation
1	A	Consecutive odd integers lie 2 units apart on the number line; so let the numbers be represented by n and n+2. Their sum is 44; so n + n + 2 = 44. 2n +2 = 44 2n = 44 - 2 2n = 42 n = 21; so n + 2 = 23
2	A	Cumulative GPA = $\frac{2.1 + 2.9 + 3.1 + n}{4}$ $3.0 = \frac{8.1 + n}{4}$ 12.0 = 8.1 + n 3.9 = n
3	C	Let m = miles/day for remaining 3 days m = (original miles - miles already traveled) / 3 $m = \frac{1924 - 490}{3} = \frac{1434}{3}$ = 478 miles per day
4	C	3x + 5 = 29 3x = 29 - 5 3x = 24 x = 8
5	D	7 - 2x = 13 -2x 7 - 13 = 0 - 6 = 0 Since this is a false statement, there is no solution to this equation.
6	B	6x + 1 = 4x - 3 6x - 4x = - 3 - 1 2x = - 4 $x = -\frac{4}{2}$ x = -2
7	D	2x + 6 + 1 = 7 + 2x 2x + 7 = 7 + 2x Since both sides of the equation are identical, any real solution satisfies the equation.

Question No.	Answer	Detailed Explanation
8	D	8x - 1 = 8x -1 = 0 Since this is a false statement, there is no solution to this equation.
9	D	2x + 5x - 9 = 8x - x - 3 - 6 7x - 9 = 7x - 9 All real values for x are solutions.
10	D	-y + 7y -54 = 0 6y - 54 = 0 6y = 54 y = 9
11	A, C & D	To solve these equations you will use your inverse operations. Depending on if it is a one step or two step you might use one or two operations. The first one is a one step equation. Since it is a multiplication problem, you will use division to undo it. In this case we are going to divide both sides by -5. When we do this, we get m = -5. The second one is also a one step equation. We divide both sides by -10. When we do this, we get c = 8. The third one is a one step equation. Since it is a addition problem, to undo the addition, we will subtract. When we subtract a -7 from both sides, we end up with g = -5. The last one is a two step equation. In this case we have to subtract 20 from both sides first. We end up with 12m = -60. Now we have to undo the multiplication. To do this we will divide both sides by 12. When we do this, we will end up with m = -5.
12	-8	To solve this 2-step equation you must first get rid of the 20. Since it is adding 20, to undo it you must subtract 20 from both sides. Making it now 5x=-40. Now the second step is to undo the multiplication. To undo it, we will divide both sides by 5. Ending up with x=-8.
13	missing no. = 13	Since you are already given the value of x, the first thing you want to do is substitute the value in. When you do this you end up with 7+(-9)=(-15)+?. Next combine your terms so you end up with -2 = -15 + ?. Next you will move the -15 to the other side. To do this you will add 15 to both sides. You end up with the unknown being 13.

Lesson 8: Solve Linear Equations with Rational Numbers

Question No.	Answer	Detailed Explanation
1	C	$\frac{7}{14} = n + (\frac{7}{14})(n)$ $\frac{1}{2} = n + (\frac{1}{2})n$; Multiply by 2 to eliminate denominators. $1 = 2n + n$ $1 = 3n$ $\frac{1}{3} = n$
2	B	$2(2x - 7) = 14$ $4x - 14 = 14$ $4x = 14 + 14$ $4x = 28$ $x = 7$
3	D	$6x - (2x + 5) = 11$ $6x - 2x - 5 = 11$ $4x - 5 = 11$ $4x = 11 + 5$ $4x = 16$ $x = 4$
4	B	$4x + 2(x - 3) = 0$ $4x + 2x - 6 = 0$ $6x - 6 = 0$ $6x = 6$ $x = 6 \div 6$ $x = 1$
5	A	$3y - 7(y + 5) = y - 35$ $3y - 7y - 35 = y - 35$ $-4y - 35 = y - 35$ $-35 = y - 35 + 4y$ $-35 + 35 = y + 4y$ $0 = 5y$ $0 = y$
6	A	$2(x-5) = \frac{1}{2}(6x + 4)$ $2x - 10 = 3x + 2$ $-12 = x$

Question No.	Answer	Detailed Explanation
7	D	$3x + 2 + x = \frac{1}{3}(12x + 6)$ $3x + x + 2 = 4x + 2$ $4x + 2 = 4x + 2$ $0 = 0$ Thus, all real values of x are correct solutions.
8	C	$\frac{1}{2}x + \frac{2}{3}x + 5 = \frac{5}{2}x + 6$ $6(\frac{1}{2}x + \frac{2}{3}x + 5 = \frac{5}{2}x + 6)$ $3x + 4x + 30 = 15x + 36$ $-8x = 6$ $x = -\frac{6}{8} = -\frac{3}{4}$
9	B	$2(2x + 3) = 3(2x + 5)$ $4x + 6 = 6x + 15$ $-2x + 6 = 15$ $-2x = 9$ $x = -\frac{9}{2}$
10	B	$.64x - .15x + .08 = .09x$ $.49x + .08 = .09x$ $.08 = -.04x$ $-.2 = x$
11	A, C	A and C are correct options. You will solve these equations the same way you did when working with integers. To start with the first one you need to get the variable alone so you add $\frac{2}{5}$ to both sides. When you do this you are left with $w=\frac{10}{5}$, which simplifies to w=2. The third one is still dealing with rational numbers, but this time in the form of decimals. First step is to get the variables on one side and the numbers on the other. So, first you will add 1.2 to both sides. You get $0.4x = 0.15x + 2$. Next move the variables to one side. You will subtract 0.15x from both sides. This will leave you with $0.25x = 2$. Last step is to divide both sides by 0.25 (or multiply by 4), and you are left with $x = 8$.

Question No.	Answer	Detailed Explanation
12	B, C & A	The equation starts off like . $\frac{3}{4} + \frac{1}{2}(m+\frac{1}{4}) = \frac{19}{16}$ The first thing you have to do is get rid of the ¾. from the left hand side of the equation. To do this you will subtract ¾ from both sides. On the right hand side you find a common denominator before you subtract. Then you will need to get rid of the ½. To do this you can either divide both sides by ½ or multiply by 2. Lastly, you need to get rid of the ¼. To do this you will subtract ¼ from each side. This will finally reveal your answer.
13	$n = \frac{1}{3}$	$\frac{8}{16} = n + \frac{8}{16}n$ Multiplying both sides by 2, we get, $2 \times \frac{8}{16} == 2\ (n + \frac{8}{16}n)$ 1 = 2n+n 3n = 1 Therefore, $n = \frac{1}{3}$

Chapter 3
Geometry & Measurement

Lesson 1: Verifying the Pythagorean Theorem

1. Which of the following could be the lengths of the sides of a right triangle?

Ⓐ 1, 2, 3
Ⓑ 2, 3, 4
Ⓒ 3, 4, 5
Ⓓ 4, 5, 6

2. A triangle has sides 8 cm long and 15 cm long, with a 90° angle between them. What is the length of the third side?

Ⓐ 7 cm
Ⓑ 17 cm
Ⓒ 23 cm
Ⓓ 289 cm

3. Find the value of c, rounded to the nearest tenth.

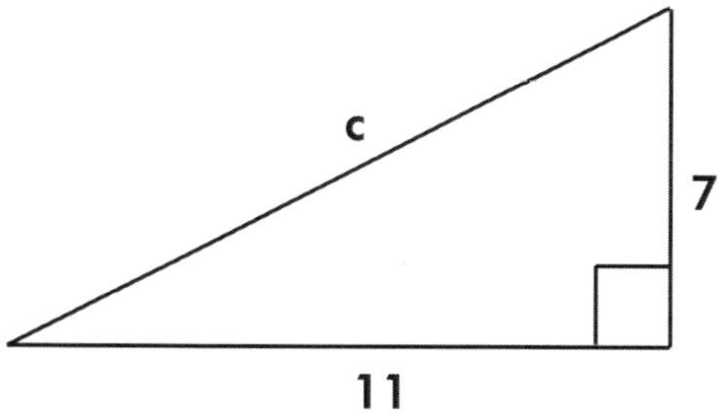

Ⓐ 8.5
Ⓑ 8.8
Ⓒ 13.0
Ⓓ 19.3

4. A square has sides 5 inches long. What is the approximate length of a diagonal of the square?

Ⓐ 5 inches
Ⓑ 6 inches
Ⓒ 7 inches
Ⓓ 8 inches

5. Which of the following INCORRECTLY completes this statement of the Pythagorean theorem?
In a right triangle with legs of lengths a and b and hypotenuse of length c, ...

Ⓐ $a^2 + b^2 = c^2$
Ⓑ $c^2 - a^2 = b^2$
Ⓒ $c^2 - b^2 = a^2$
Ⓓ $a^2 + c^2 = b^2$

6. A Pythagorean triplet is a set of three positive integers a, b, and c that satisfy the equation $a^2 + b^2 = c^2$. Which of the following is a Pythagorean triple?

Ⓐ a = 3, b = 6, c = 9
Ⓑ a = 6, b = 9, c = 12
Ⓒ a = 9, b = 12, c = 15
Ⓓ a = 12, b = 15, c = 18

7. If an isosceles right triangle has legs of 4 inches each, find the length of the hypotenuse.

Ⓐ Approximately 6 in.
Ⓑ Approximately 5 in.
Ⓒ Approximately 4 in.
Ⓓ Approximately 3 in.

8. In triangle ABC, angle C = 90°, AC = 4 and AB = 10. Find BC to the nearest tenth.

Ⓐ 9.5
Ⓑ 9.2
Ⓒ 8.9
Ⓓ 8.5

9. In triangle ABC, angle C = 90°, AB = 35, and BC = 28. Find AC.

Ⓐ 23
Ⓑ 22
Ⓒ 21
Ⓓ 20

10. The diagonal of a square is 25. Find the approximate side lengths.

Ⓐ 15
Ⓑ 16
Ⓒ 17
Ⓓ 18

11. You have a right triangle with leg lengths of 6 and 8. What is the length of the hypotenuse? Fill in the numbers into the equation and solve.

$6^2 + 8^2 = C^2$
$100 = C^2$
$\sqrt{100} = C$
$C=?$

12. Which equation would you use to solve for the missing side of the triangle pictured below?

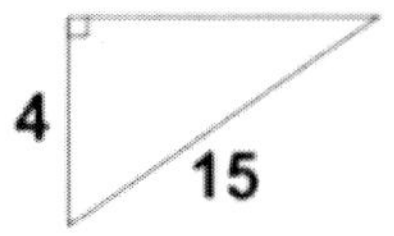

Ⓐ $4^2+15^2=x^2$
Ⓑ $4^2+x^2=15^2$
Ⓒ $x^2+15^2=4^2$

CHAPTER 3 → Lesson 2: Pythagorean Theorem in Real-World Problems

1. The bottom of a 17-foot ladder is placed on level ground 8 feet from the side of a house as shown in the figure below. Find the vertical height at which the top of the ladder touches the side of the house.

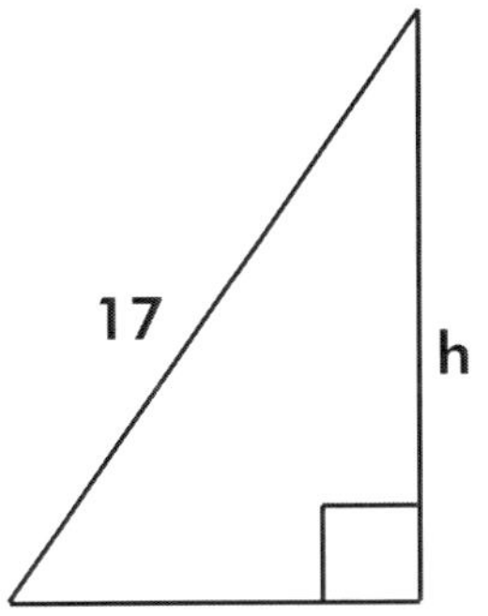

Ⓐ h = 9 feet
Ⓑ h = 12 feet
Ⓒ h = 15 feet
Ⓓ h = 18 feet

2. Which of the following equations could be used to find the value of w?

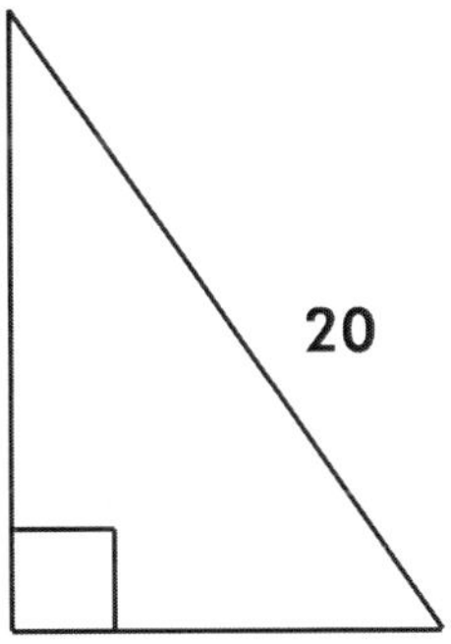

Ⓐ $w^2 + 18^2 = 20^2$
Ⓑ $18^2 - w^2 = 20^2$
Ⓒ $20^2 + 18^2 = w^2$
Ⓓ $w + 18 = 20$

3. John has a chest where he keeps his antiques. What is the measure of the diagonal (d) of John's chest with the height (c) = 3ft, width (b) = 3ft, and length (a) = 5ft.?

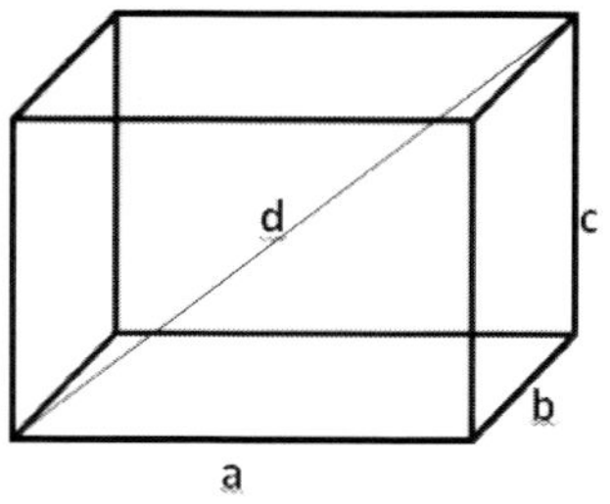

Ⓐ $\sqrt{42}$ ft²
Ⓑ $\sqrt{44}$ ft²
Ⓒ $\sqrt{34}$ ft
Ⓓ $\sqrt{43}$ ft

4. Mary has a lawn that has a width (a) of 30 yards, and a length (b) of 40 yards. What is the measurement of the diagonal (c) of the lawn?

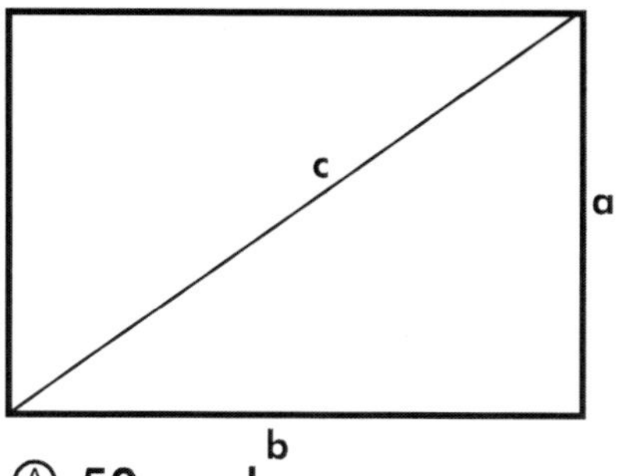

Ⓐ 50 yards
Ⓑ 49 yards
Ⓒ 51 yards
Ⓓ 50 yards²

5. A construction company needed to build a sign with the width (a) of 9 ft, and a length (b) of 20 ft. What will be the approximate measurement of the diagonal (c) of the sign?

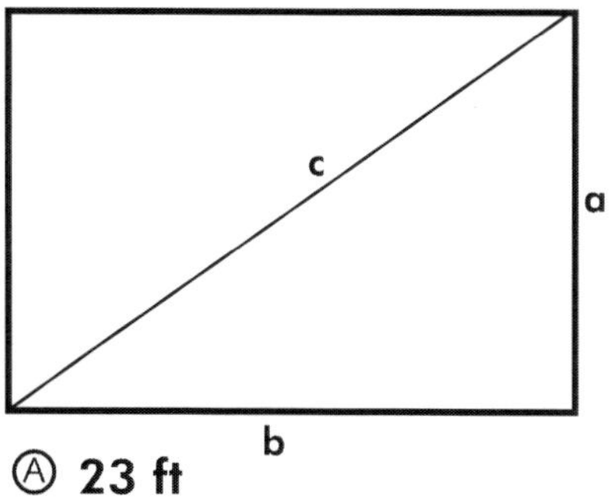

Ⓐ 23 ft
Ⓑ 22 ft
Ⓒ 20 ft
Ⓓ 19 ft

6. An unofficial baseball diamond is measured to be 50 yards wide. What is the approximate measurement of one side (a) of the diamond?

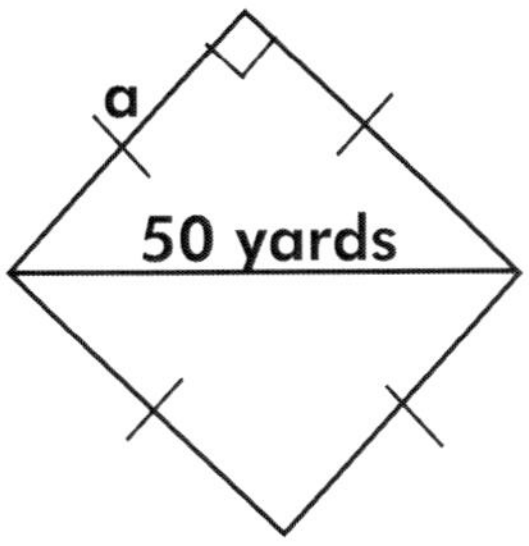

Ⓐ 34 ft
Ⓑ 35 yards
Ⓒ 35 ft
Ⓓ 36 yards

7. The neighborhood swimming pool is 20 ft wide and 30 ft long. What is the approximate measurement of the diagonal (d) of the base of the pool?

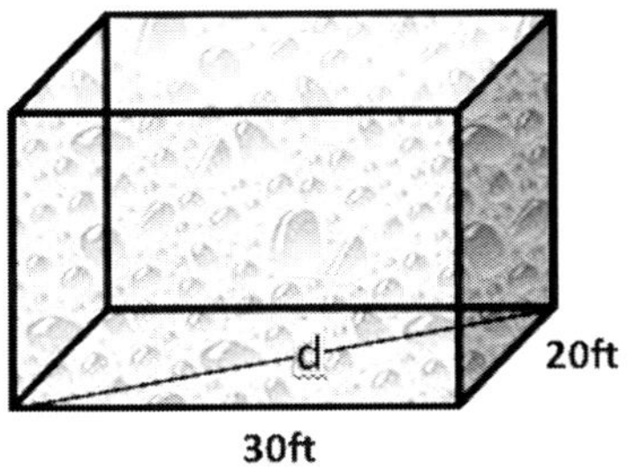

Ⓐ 36 ft
Ⓑ 35 ft
Ⓒ 34 ft
Ⓓ 34 yards

8. To get to her friend's house, a student must walk 20 feet to the corner of their streets, turn left and walk 15 feet to her friend's house.
How much shorter would it be if she could cut across a neighbor's yard and walk a straight line from her house to her friend's house?

Ⓐ 5 feet shorter
Ⓑ 10 feet shorter
Ⓒ 25 feet shorter
Ⓓ 35 feet shorter

9. Your school principal wants the custodian to put a new flag up on the flagpole. If the flagpole is 40 feet tall and they have a 50 foot ladder, approximately how far from the base of the pole can he place the base of his ladder in order to accomplish the task?

Ⓐ Up to 10 ft away
Ⓑ Up to 20 ft away
Ⓒ Up to 30 ft away
Ⓓ Up to 40 ft away

10. Your kite is at the end of a 50 ft string. You are 25 ft from the outside wall of a building that you know makes a right angle with the ground.
How high is the kite approximately?

Ⓐ Approximately 39 ft high
Ⓑ Approximately 40 ft high
Ⓒ Approximately 42 ft high
Ⓓ Approximately 43 ft high

11. Match the following word problem with the correct equation that you would use to solve it.

	$9^2+x^2=15^2$	$40^2+38^2=x^2$	$9^2+15^2=x^2$	$x^2+38^2=40^2$
One house is 15 miles due north of the park. Another house is 9 miles due east of the park. How far apart are the houses from each other?	○	○	○	○
The foot of a ladder is put 9 feet from the wall. If the ladder is 15 feet long how high up the building will the ladder reach?	○	○	○	○
If you drive your car 40 miles south and then 38 miles east, how far would the shortest route be from your starting point?	○	○	○	○
The diagonal of a TV is 40 inches. The TV is 38 inches long. How tall is the TV?	○	○	○	○

12. Fill in the missing information needed to solve for the word problem. Round to the nearest tenth if necessary.

WORD PROBLEM	LEG(a)	LEG(b)	HYPOTENUSE(c)
Find the height of a pyramid whose slant height is 26 cm and base length is 48 cm	24		
Find the base length of a pyramid whose height is 8 in and slant height 17 in.		8	17
The foot of a ladder is put 5 feet from the wall. If the top of the ladder is 10 feet from the ground, how long is the ladder?		10	

13. On a bike ride you start at your house and ride your bike 3.5 miles north and then 1.25 miles east. How far are you directly from your house? Select the correct equation with the correct solution to match the word problem.

Ⓐ $\sqrt{3.5^2+1.25^2}$ = 22.56 mi
Ⓑ $\sqrt{3.5^2-1.25^2}$ = 3.34 mi
Ⓒ $\sqrt{3.5^2+1.25^2}$ = 3.72 mi

CHAPTER 3 → Lesson 3: Pythagorean Theorem & Coordinate System

1. A robot begins at point A, travels 4 meters west, then turns and travels 7 meters south, reaching point B. What is the approximate straight-line distance between points A and B?

Ⓐ 8 meters
Ⓑ 9 meters
Ⓒ 10 meters
Ⓓ 11 meters

2. What is the distance between the points (1, 3) and (9, 9)?

Ⓐ 6 units
Ⓑ 8 units
Ⓒ 10 units
Ⓓ 12 units

3. Find the distance (approximately) between Pt A (2, 7) and Pt B (-2, -7).

Ⓐ 14.0
Ⓑ 14.6
Ⓒ 18.0
Ⓓ 13.4

4. Find the distance (approximately) between Pt P (5, 3) and the origin (0, 0).

Ⓐ 4.0
Ⓑ 5.1
Ⓒ 5.8
Ⓓ 8.0

5. Find the distance (approximately) between the points A (11, 12) and B (7, 8).

Ⓐ 4.0
Ⓑ 5.3
Ⓒ 5.7
Ⓓ 8.0

6. Is it closer to go from Pt A (4, 6) to Pt B (2, -4) or Pt A to Pt C (-5, 2)?

Ⓐ A to B
Ⓑ A to C
Ⓒ Neither, they are both the same distance.
Ⓓ Not enough information.

7. Paul lives 50 yards east and 40 yards south of his friend, Larry. If he wants to shorten his walk by walking in a straight line from his home to Larry's, how far (approximately) will he walk?

Ⓐ 64 yards
Ⓑ 62 yards
Ⓒ 60 yards
Ⓓ 58 yards

8. In a coordinate plane, a point P (7,8) is rotated 90° clockwise around the origin and then reflected across the vertical axis. Find the distance (approximately) between the original point and the final point.

Ⓐ 14 units
Ⓑ 21 units
Ⓒ 24 units
Ⓓ None of the above

9. You are using a coordinate plane to sketch out a plan for your vegetable garden. Your garden will be a rectangle 15 ft wide and 20 ft long. You want a square in the center with 3 ft sides to be reserved for flowers. If the garden is plotted on the coordinate plane with the southwest corner at the origin, what are the coordinates of the center of the flower garden? (Assume that the longer side is placed on the y-axis)

Ⓐ (15, 10)
Ⓑ (7.5, 10)
Ⓒ (-15, 10)
Ⓓ (-7.5, 10)

10. You are using a coordinate plane to sketch out a plan for a vegetable garden. The garden will be a rectangle 15 feet wide and 20 feet long. You want a square in the center with 3 feet sides to be reserved for flowers. The garden is plotted in the coordinate grid so that the southwest corner is placed at the origin. Find the length of the diagonal (approximately) of the flower bed.

Ⓐ 4.0 feet
Ⓑ 4.2 feet
Ⓒ 5.0 feet
Ⓓ 5.2 feet

11. Match the ordered pairs with the approximate distance between them.

	10.8	12.2	13	14.8
(6, 5) and (-4, 9)	○	○	○	○
(-8, 0) and (5, -7)	○	○	○	○
(-4, -9) and (6, -2)	○	○	○	○
(5, 4) and (12, 15)	○	○	○	○

12. Given below are the coordinates of the hypotenuse of right-angled triangles. Fill in the missing values with the lengths of the sides. Round to the nearest hundredth if necessary.

ORDERED PAIRS	LEG LENGTH	LEG LENGTH	HYPOTENUSE
(6,2),(0,-6)	6		
(-3,-1),(-4,0)	1	1	
(-2,3),(-1,7)			4.12

CHAPTER 3 → Lesson 4: Finding Volume: Cone, Cylinder, & Sphere

1. What is the volume of a sphere with a radius of 6?

Ⓐ 72π
Ⓑ 144π
Ⓒ 216π
Ⓓ 288π

2. A cone has a height of 9 and a base whose radius is 4. Find the volume of the cone.

Ⓐ 18π
Ⓑ 36π
Ⓒ 48π
Ⓓ 72π

3. What is the volume of a cylinder with a radius of 5 and a height of 3?

Ⓐ 30π
Ⓑ 45π
Ⓒ 75π
Ⓓ 120π

4. A round ball has a diameter of 10 inches. What is the approximate volume of the ball? Use $\pi = 3.14$

Ⓐ 130 cubic inches
Ⓑ 260 cubic inches
Ⓒ 390 cubic inches
Ⓓ 520 cubic inches

5. A cylindrical can has a height of 5 inches and a diameter of 4 inches. What is the approximate volume of the can? Use $\pi = 3.14$

Ⓐ 20 cubic inches
Ⓑ 60 cubic inches
Ⓒ 100 cubic inches
Ⓓ 200 cubic inches

6. A cylinder and a cone have the same radius and the same volume. How do the heights compare?

Ⓐ The height of the cylinder is 3 times the height of the cone.
Ⓑ The height of the cylinder is 2 times the height of the cone.
Ⓒ The height of the cone is 2 times the height of the cylinder.
Ⓓ The height of the cone is 3 times the height of the cylinder.

7. Which of the following has the greatest volume?

Ⓐ A sphere with a radius of 2 cm
Ⓑ A cylinder with a height of 2 cm and a radius of 2 cm
Ⓒ A cone with a height of 4 cm and a radius of 3 cm
Ⓓ All three volumes are equal

8. Which of the following has the greatest volume?

Ⓐ A sphere with a radius of 3 cm
Ⓑ A cylinder with a height of 4 cm and a radius of 3 cm
Ⓒ A cone with a height of 3 cm and a radius of 6 cm
Ⓓ All three volumes are equal

9. If a sphere and a cone have the same radii r and the cone has a height of 4, find the ratio of the volume of the sphere to the volume of the cone.

Ⓐ r : 1
Ⓑ 1 : r
Ⓒ r : 4
Ⓓ 4 : r

10. Which of the following describes the relationship between the volumes of a cone and cylinder with the same radii and the same height.

Ⓐ The volume of the cylinder is three times that of the cone.
Ⓑ The volume of the cylinder is 1/3 that of the cone.
Ⓒ The volume of the cylinder is 4/3 that of the cone.
Ⓓ None of the above.

11. Find the volume of the figure below. Use pi = 3.14. Write your answer in the box below.

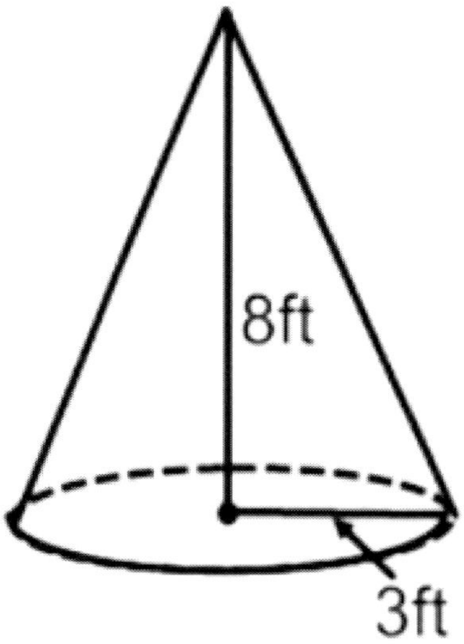

12. Select the equation you would use to find the volume of the sphere pictured.

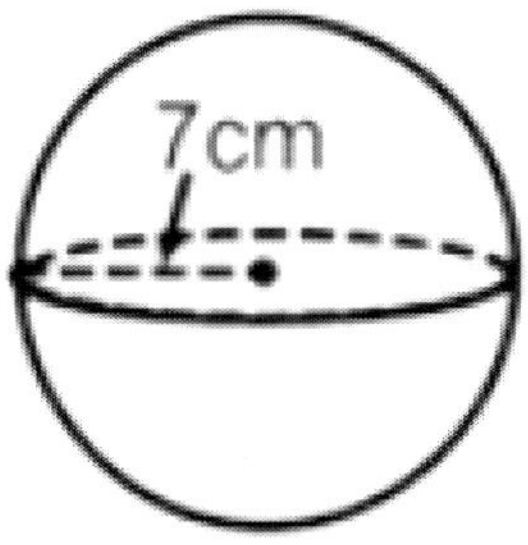

Ⓐ $4 \times \pi \times 7^2$

Ⓑ $\frac{4}{3} \times \pi \times 7^3$

Ⓒ $\frac{4}{3} \times \pi \times 7^2$

End of Geometry & Measurement

ANSWER KEY AND DETAILED EXPLANATION

Chapter 3
Geometry & Measurement

Lesson 1: Verifying the Pythagorean Theorem

Question No.	Answer	Detailed Explanation
1	C	$3^2+4^2=5^2$ $9+16=25$ $25=25$
2	B	$8^2+15^2=c^2$ $64+225=c^2$ $289=c^2$ 17 cm $= c$
3	C	$7^2+11^2=c^2$ $49+121=c^2$ $170=c^2$ $13.0\approx c$
4	C	$5^2+5^2=c^2$ $25+25=c^2$ $50=c^2$ 7 inches$\approx c$
5	D	$a^2+b^2=c^2$ cannot be changed to $a^2+c^2=b^2$.
6	C	$9^2+12^2=15^2$ $81+144=225$ $225=225$
7	A	$4^2+4^2= c^2$ $32 = c^2$ 5.7 in $\approx c$
8	B	$16+a^2 = 100$ $a^2 = 84$ $a \approx 9.2$

Question No.	Answer	Detailed Explanation
9	C	$AC^2+BC^2=AB^2$ $AC^2+28^2=35^2$ $AC^2=1225-784$ $AC^2=441$ $AC=21$
10	D	$2s^2=625$ $s^2=312.5$ s is approximately 18
11	10	You will fill the legs into the equation. So the equation will be $6^2 + 8^2 = c^2$. When you square both of those you get $100 = c^2$. To get c alone you have to take the square root of both sides. When you do this you have $\sqrt{100} = C$. So once you take the square root, you are left with c = 10.
12	B	In order to solve this equation, you need to have 15 as the hypotenuse. This means that it is one the opposite side of the equal sign. The equation you should select is $4^2 + x^2 = 15^2$.

Lesson 2: Pythagorean Theorem in Real-World Problems

Question No.	Answer	Detailed Explanation
1	C	$8^2+h^2=17^2$ $h^2=289-64$ $h^2=225$ h=15 feet
2	A	Applying the Pythagorean Theorem to this (w, 18, 20) right triangle, $w^2 + 18^2 = 20^2$ is the correct equation.
3	D	5^2+3^2=(diagonal of bottom of chest)^2 $\sqrt{34}$=diagonal of bottom of chest $34+9=d^2$ $43=d^2$ $\sqrt{43}$ ft =d
4	A	Remember the Pythagorean triples (3,4,5 and their multiples). In this case: 30, 40, 50
5	B	$9^2+20^2=c^2$ $81+400=c^2$ $481=c^2$ $\sqrt{481}=c$ c is approximately 22 ft
6	B	$a^2+a^2=50^2$ $2a^2=2500$ $a^2=1250$ a=35 yd (approx)
7	A	$30^2+20^2=d^2$ $1300=d^2$ $\sqrt{1300}=d$ d=approx. 36 ft
8	B	$20^2+15^2=d^2$ $625=d^2$ 25 ft=d 20+15=35 ft The diagonal would be 10 ft shorter.
9	C	Use the Pythagorean triple: 30, 40, 50 He can place his ladder up to 30 feet away.

<table>
<tr><th>Question No.</th><th>Answer</th><th>Detailed Explanation</th></tr>
<tr><td>10</td><td>D</td><td>$25^2+x^2=50^2$
$x^2=2500-625=1875$
x=43 ft (approx)
Alternate Method :
$x^2 = 50^2 - 25^2$
$x^2 = 25^2 \times 2^2 - 25^2$
$x^2 = 25^2 (2^2 - 1)$
$x = \sqrt{(25^2 \times 3)}$
$x = 25 \times \sqrt{3}$
x = 25 x 1.73 (when $\sqrt{3}$ is rounded to the nearest hundredth)
x = 43 ft (approx)</td></tr>
<tr><td>11</td><td></td><td>
<table>
<tr><th></th><th>$9^2+x^2=15^2$</th><th>$40^2+38^2=x^2$</th><th>$9^2+15^2=x^2$</th><th>$x^2+38^2=40^2$</th></tr>
<tr><td>One house is 15 miles due north of the park. Another house is 9 miles due east of the park. How far apart are the houses from each other?</td><td></td><td></td><td>●</td><td></td></tr>
<tr><td>The foot of a ladder is put 9 feet from the wall. If the ladder is 15 feet long how high up the building will the ladder reach?</td><td>●</td><td></td><td></td><td></td></tr>
<tr><td>If you drive your car 40 miles south and then 38 miles east, how far would the shortest route be from your starting point?</td><td></td><td>●</td><td></td><td></td></tr>
<tr><td>The diagonal of a TV is 40 inches. The TV is 38 inches long. How tall is the TV?</td><td></td><td></td><td></td><td>●</td></tr>
</table>
The first story gives you the 2 legs of the triangles. So you will need to use the equation that will solve for the hypotenuse. The second one, the ladder is your hypotenuse, so you will be solving for one of the legs. The third one, they give you the legs, so you are solving for the hypotenuse. The last one, the diagonal is the hypotenuse, so you are solving for a leg.</td></tr>
</table>

Question No.	Answer	Detailed Explanation
12	(1) 10, 26 (2) 30 (3) 5, 11.2	(1) The first example you are looking for the other leg. Since the base length is 48, you have to cut it in half to make the right triangle. So in this case, one of the legs will be 24 cm. The hypotenuse is going to be the slant height of 26 cm. This means you are solving for the other leg. When you fill the numbers into Pythagorean Theorem you get 10 cm. (2) The next example you have both a leg and a hypotenuse, so you are solving for the other leg. In this case the other leg will be 15 in. which is half of the base length. So, base length will be 30 in. (3) The last example you have both legs, 5 ft. and 10 ft. So the length of the ladder will be the hypotenuse. In this case it will be about 11.2 ft.
13	C	The 2 directions you biked formed the legs of the triangle. So you are solving for the hypotenuse. When you solve this you will take the two legs and square them and add them together. The last step is to take the square root of that sum. In this case, third choice is the correct answer.

Lesson 3: Pythagorean Theorem & Coordinate System

Question No.	Answer	Detailed Explanation
1	A	$4^2+7^2=d^2$ $65=d^2$; d is approx. 8 m
2	C	Draw the graph. Draw the right triangle with the distance between the points as the hypotenuse. The legs will be 6 units & 8 units. Find the hypotenuse. $8^2+6^2=h^2$ $64+36=h^2$ $100=h^2$; $10=h$
3	B	$4^2+14^2=d^2$ $16+196=d^2$ $212=d^2$ $d=14.6$ (approximately)
4	C	$25+9=34=d^2$ $d=5.8$ (approximately)
5	C	$4^2+4^2=d^2$ $32=d^2$ $d=5.7$ (approximately)
6	B	$2^2+10^2=AB^2$ $\sqrt{104}=AB$ $10.2=AB$ (approximately) $4^2+9^2=AC^2$ $\sqrt{97}=AC$ $9.8=AC$ (approximately) Note that, we can conclude $AB > AC$ without calculating square root also, because $AB^2>AC^2$ implies $AB > AC$
7	A	$50^2+40^2=d^2$ $2500+1600=d^2$ $4100=d^2$ 64 yd is approx. distance
8	B	Original point: (7,8) New point: (8,-7) Final point: (-8,-7) $15^2+15^2=d^2$ $450=d^2$ $d=21.2$ units (approximately)

<table>
<tr><th>Question No.</th><th>Answer</th><th>Detailed Explanation</th></tr>
<tr><td>9</td><td>B</td><td>Draw a sketch. The coordinates of the rectangle are (0,0), (15,0), (15,20), and (0,20).
Halfway across the width is 7.5 and halfway up the length is 10.
Therefore, the center is at (7.5,10).</td></tr>
<tr><td>10</td><td>B</td><td>$3^2+3^2=d^2$
$18=d^2$; $4.2=d$ (approximately)</td></tr>
<tr><td>11</td><td></td><td><table>
<tr><th></th><th>10.8</th><th>12.2</th><th>13</th><th>14.8</th></tr>
<tr><td>(6, 5) and (-4, 9)</td><td>●</td><td></td><td></td><td></td></tr>
<tr><td>(-8, 0) and (5, -7)</td><td></td><td></td><td></td><td>●</td></tr>
<tr><td>(-4, -9) and (6, -2)</td><td></td><td>●</td><td></td><td></td></tr>
<tr><td>(5, 4) and (12, 15)</td><td></td><td></td><td>●</td><td></td></tr>
</table>
Distance between two points having coordinates (x_1,y_1) and (x_2,y_2) can be found out with the help of Pythagorean Theorem. $D=\sqrt{(x_2-x_1)^2+(y_2-y_1)^2}$. Using this formula we get -
(i) d = 10.8
(ii) d = 14.8
(iii) d = 12.2
(iv) d = 13</td></tr>
<tr><td>12</td><td>8||10||1.41||1||4</td><td>The first set of ordered pairs have the missing leg length of 8 which we can then fill those lengths into Pythagorean Theorem. So the hypotenuse is 10. The second set you already were given both leg lengths so they need to be filled into Pythagorean Theorem. This will make the hypotenuse 1.41 (approximately). The last set already gives you the hypotenuse so you just need to plot the ordered pairs and form your triangle. When you do this, you will see that the legs lengths are 1 and 4.</td></tr>
</table>

Lesson 4: Finding Volume: Cone, Cylinder, & Sphere

Question No.	Answer	Detailed Explanation
1	D	$V=\frac{4}{3}(\pi r^3)$ $V=\frac{4}{3}(\pi)\ 6^3$ $V=\frac{4}{3}(216)\ \pi$ $V=288\pi$
2	C	$V=(\frac{1}{3})Bh$ where B= area of the base $V=(\frac{1}{3})\pi(4^2)(9)$ $V=48\pi$
3	C	$V=\pi r^2h$ $V=\pi\ (25)(3)$ $V=75\pi$
4	D	$V=(\frac{4}{3})\pi r^3$ $V=\frac{4}{3}(125\pi)$ $V=(\frac{500}{3})\pi$ V=523.6 cubic inches
5	B	$V=\pi r^2h$ $V=(5)(2)^2\pi$ $V=20\pi$ V=20(3.14) cu in V=62.8 cu in = 60 cu in (when rounded to the nearest ten)
6	D	$V_{cylinder} = \pi r^2h$ $V_{cone}=\frac{1}{3}\pi r^2h$ If volumes are the same, then the height of the cone must be 3 times the height of the cylinder because $3(\frac{1}{3})=1$.

Question No.	Answer	Detailed Explanation
7	C	Vsphere=$\frac{4}{3}\pi r^3$ V=$\frac{4}{3}(8)\pi$ V=$\frac{32}{3}\pi$ Vcylinder=$\pi r^2 h$ V=8π Vcone=$\frac{1}{3}\pi r^2 h$ V=$(\frac{1}{3})(9)(4)\pi$ V=12π; The cone has the greatest volume.
8	D	Vsphere=$\frac{4}{3}\pi r^3$ V=$\frac{4}{3}\pi(27)$ V=36π Vcylinder=$\pi r^2 h$ V=$9(4)\pi$ V=36π Vcone=$\frac{1}{3}\pi r^2 h$ V=$(\frac{1}{3})(36)(3)\pi$ V=36π
9	A	Vsphere/Vcone = $(\frac{4}{3})\pi r^3 / (\frac{1}{3})\pi r^2 h = 4r/h = 4r/4 = r/1$
10	A	Vcylinder/Vcone= $\pi r^2 h/(\frac{1}{3})\pi r^2 h=3/1$
11	75.36 ft^3	Radius of the cone = r = 3 ft. Height of the cone = h = 8 ft. Volume of the cone = V= $(1/3)\pi r^2 h$ V = $(1/3) \times 3.14 \times 3^2 \times 8$ V = 75.36 ft^3
12	B	The formula used to find the volume of a sphere is V= $(\frac{4}{3})\pi r^3$. So you just have to fill in the radius of 7 into the equation for r. Thus making it the second choice.

Date of Completion:______________ Score:______________

Chapter 4
Data & Probability

Lesson 1: Interpreting Data Tables & Scatter Plots

1. If a scatter plot has a line of best fit that decreases from left to right, which of the following terms describes the association?

Ⓐ Positive association
Ⓑ Negative association
Ⓒ Constant association
Ⓓ Nonlinear association

2. If a scatter plot has a line of best fit that increases from left to right, which of the following terms describes the association?

Ⓐ Positive association
Ⓑ Negative association
Ⓒ Constant association
Ⓓ Nonlinear association

3. Which of the following scatter plots is the best example of a linear association?

Ⓐ
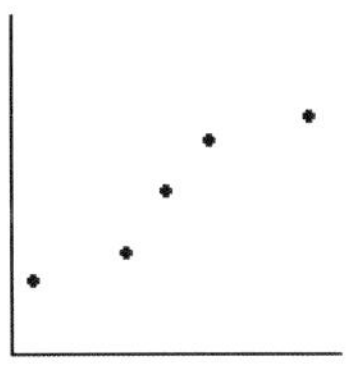

Ⓑ
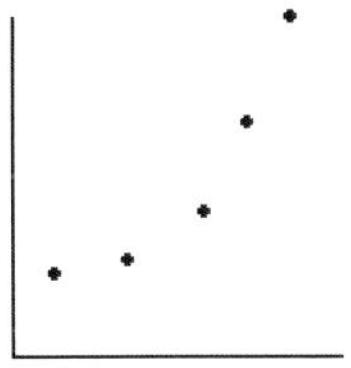

Ⓒ
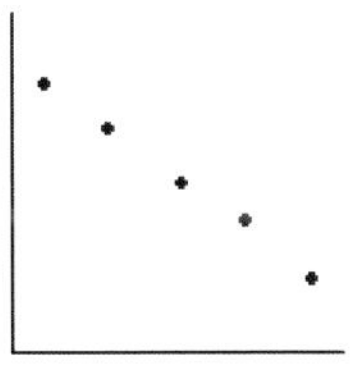

Ⓓ

4. Data for 9 kids' History and English grades are made available in the chart. What is the association between the History and English grades?

Kids	1	2	3	4	5	6	7	8	9
History	63	49	84	33	55	23	71	62	41
English	67	69	82	32	59	26	73	62	39

Ⓐ Positive association
Ⓑ Negative association
Ⓒ Nonlinear association
Ⓓ Constant association

5. Data for 9 kids' History grades and the distance they live from school are made available in the chart. What is the association between these two categories?

Kids	1	2	3	4	5	6	7	8	9
History	63	49	84	33	55	23	71	62	41
Distance from School (miles)	.5	7	3	4	5	2	3	6	9

Ⓐ No association
Ⓑ Positive association
Ⓒ Negative association
Ⓓ Constant association

6. Data for 9 kids' Math and Science grades are made available in the chart. What is the association between the Math and Science grades?

Kids	1	2	3	4	5	6	7	8	9
Science	63	49	84	33	55	23	71	62	41
Math	67	69	82	32	59	26	73	62	39

Ⓐ Positive association
Ⓑ No association
Ⓒ Constant association
Ⓓ Negative association

7. Which of the scatter plots below is the best example of positive association?

Ⓐ 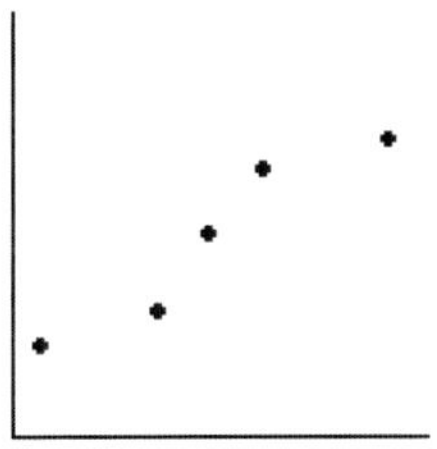

Ⓑ 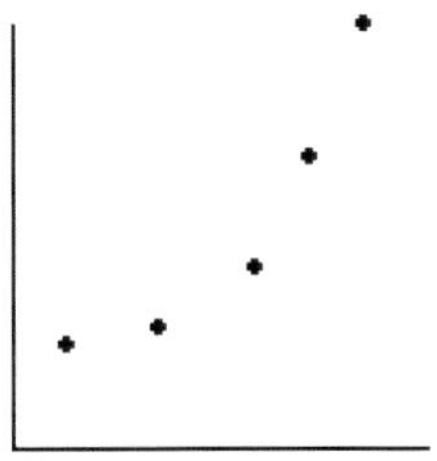

Ⓒ 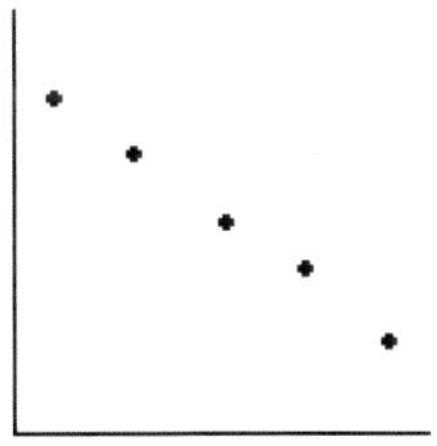

Ⓓ 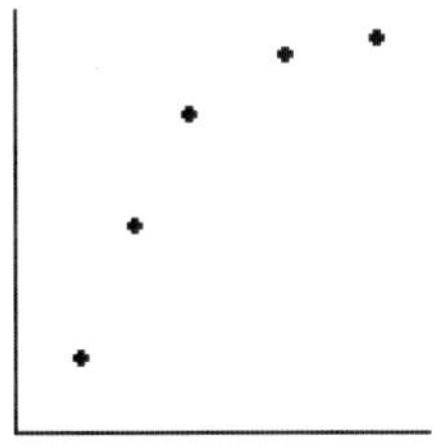

8. 150 students were surveyed and asked whether they played a sport and whether they played a musical instrument. The results are shown in the table below.

	Plays an Instrument	Does not Plays an Instrument
Plays a Sport	60	30
Does not Plays a Sport	10	50

What percent of the 150 students play a sport and also do not play an instrument?

Ⓐ 20%
Ⓑ 33%
Ⓒ 40%
Ⓓ 50%

9. 150 students were surveyed and asked whether they played a sport and whether they played a musical instrument. The results are shown in the table below.

	Plays an Instrument	Does not Plays an Instrument
Plays a Sport	60	30
Does not Plays a Sport	10	50

Which of the following statements is NOT supported by the data?

Ⓐ A randomly chosen student who plays a sport is 2 times as likely to play an instrument as to not play an instrument.
Ⓑ A randomly chosen student who does not play an instrument is 2 times as likely to not play a sport as to play a sport.
Ⓒ A randomly chosen student who does not play a sport is 5 times as likely to not play an instrument as to play an instrument.
Ⓓ A randomly chosen student who plays an instrument is 6 times as likely to play a sport as to not play a sport.

10. 150 students were surveyed and asked whether they played a sport and whether they played a musical instrument. The results are shown in the table below.

	Plays an Instrument	Does not Plays an Instrument
Plays a Sport	60	30
Does not Plays a Sport	10	50

Which two sections add up to just over half of the number of students surveyed?

Ⓐ The two sections that do not play an instrument.
Ⓑ The two sections that do not play a sport.
Ⓒ The two sections that play an instrument.
Ⓓ The two sections that play a sport.

11. Match the data with the correct association.

<table>
<tr><th></th><th>POSITIVE ASSOCIA-TION</th><th>NEGATIVE ASSOCIA-TION</th><th>NO ASSOCIATION</th></tr>
<tr><td>The population survey data for 5 years shows the number of goldfish and star fish. Describe the association between the population of goldfish and star fish.
<table>
<tr><td>YEAR</td><td>1</td><td>2</td><td>3</td><td>4</td><td>5</td></tr>
<tr><td>GOLDFISH</td><td>13</td><td>18</td><td>19</td><td>20</td><td>25</td></tr>
<tr><td>STARFISH</td><td>30</td><td>25</td><td>20</td><td>15</td><td>12</td></tr>
</table></td><td>○</td><td>○</td><td>○</td></tr>
<tr><td>Below is data for 5 years showing Jonny's and Jack's weight in kg. Describe the association between the weight of Jonny and Jack.
<table>
<tr><td>YEAR</td><td>1</td><td>2</td><td>3</td><td>4</td><td>5</td></tr>
<tr><td>JONNY</td><td>30</td><td>40</td><td>42</td><td>49</td><td>52</td></tr>
<tr><td>JACK</td><td>30</td><td>35</td><td>40</td><td>45</td><td>50</td></tr>
</table></td><td>○</td><td>○</td><td>○</td></tr>
<tr><td>The data for 5 days shows the sale of watermelon and potatoes. Describe the association between the sale of water-melon and potatoes.
<table>
<tr><td>DAYS</td><td>1</td><td>2</td><td>3</td><td>4</td><td>5</td></tr>
<tr><td>WATERMELON</td><td>70</td><td>26</td><td>60</td><td>19</td><td>70</td></tr>
<tr><td>POTATO</td><td>10</td><td>40</td><td>15</td><td>80</td><td>22</td></tr>
</table></td><td>○</td><td>○</td><td>○</td></tr>
</table>

12. Following is 10 days of data which shows the sale of apples and mangoes. Fill in the type of association there is between the apple and mango sales.

DAYS	1	2	3	4	5	6	7	8	9	10
APPLE	62	49	81	26	45	55	16	74	97	34
MANGO	36	44	49	37	26	11	76	83	64	81

There is ________________________ between apple and mango sale.

13. Fill in the table with the word positive, negative, or none to describe which type of association is plotted.

Scatter Plot	Type of Association
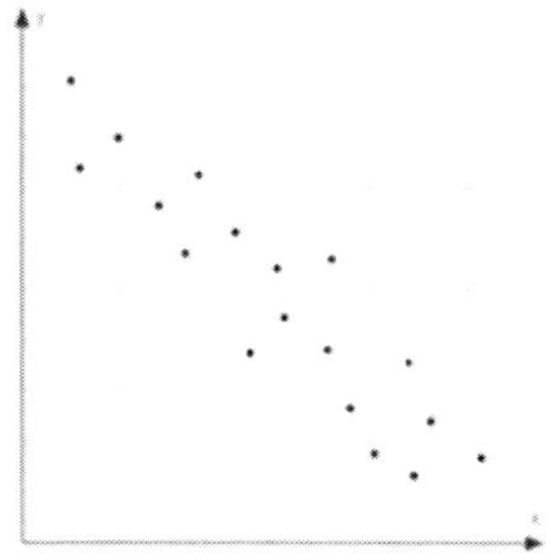	
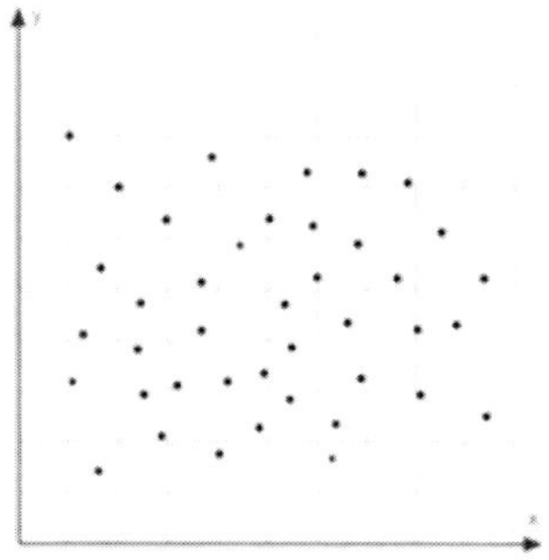	
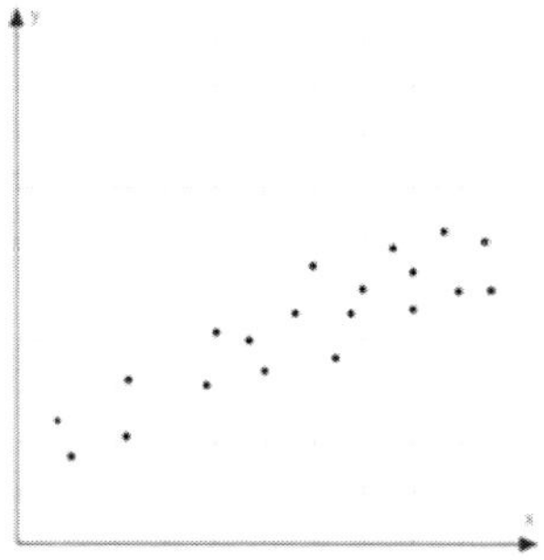	
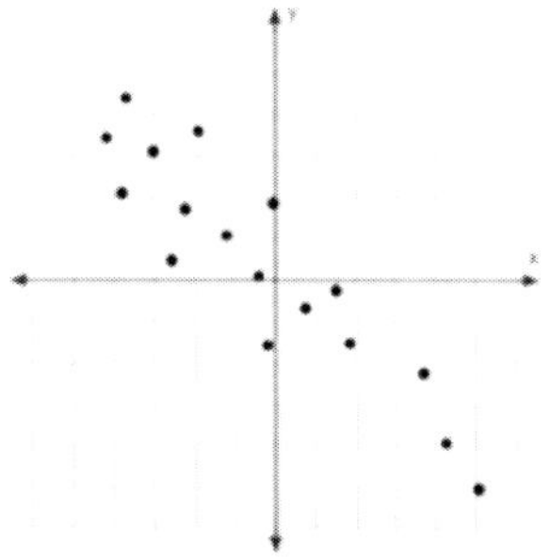	

CHAPTER 4 → Lesson 2: Scatter Plots, Line of Best Fit

1. 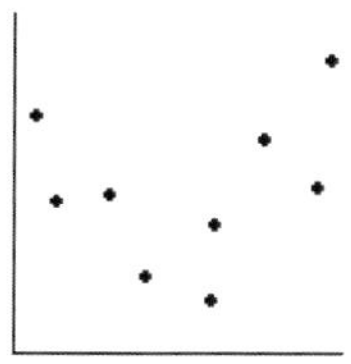

Which of the following best describes the points in this scatter plot?

Ⓐ Increasing Linear
Ⓑ Decreasing Linear
Ⓒ Constant Linear
Ⓓ None of these

2. 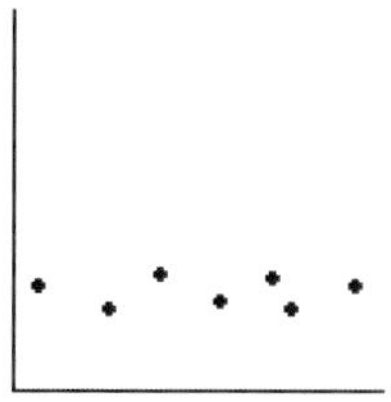

Which of the following best describes the points in this scatter plot?

Ⓐ Increasing Linear
Ⓑ Decreasing Linear
Ⓒ Constant Linear
Ⓓ None of these

3. 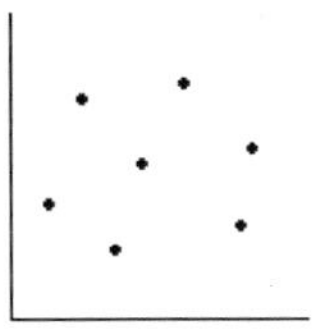

Which of the following lines best approximates the data in the scatter plot shown above?

Ⓐ 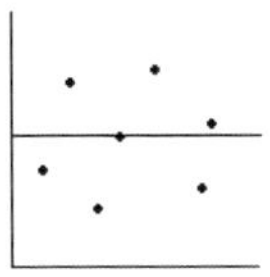

Ⓑ 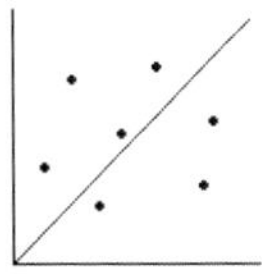

Ⓒ 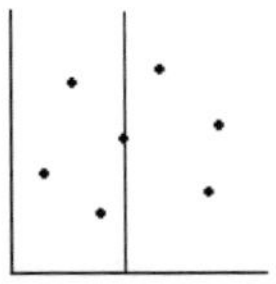

Ⓓ None of these; the data do not appear to be related linearly.

4. Which scatter plot represents a positive linear association?

Ⓐ 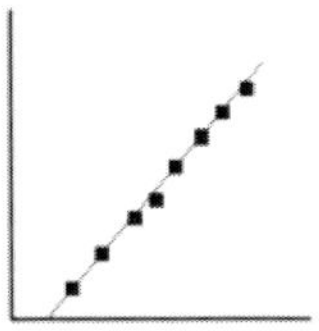

Ⓑ 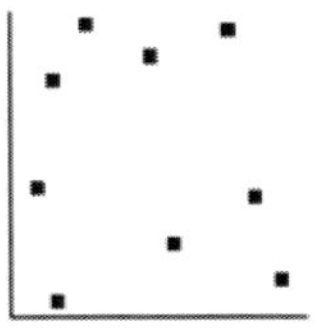

Ⓒ 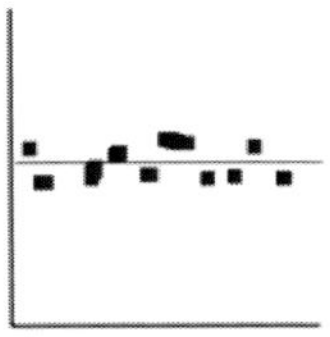

Ⓓ 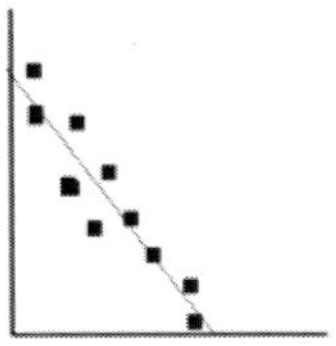

5. Which scatter plot represents a negative linear association?

Ⓐ 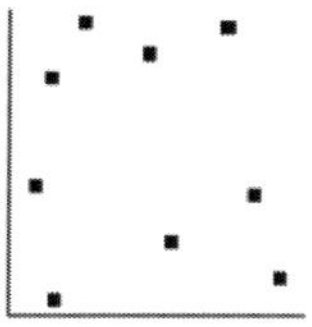

Ⓑ 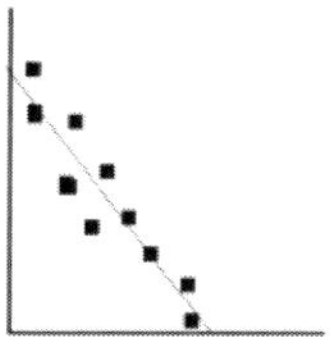

Ⓒ 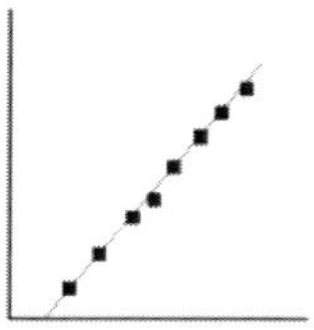

Ⓓ 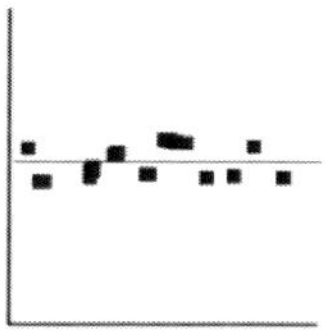

6. Which scatter plot represents no association?

Ⓐ 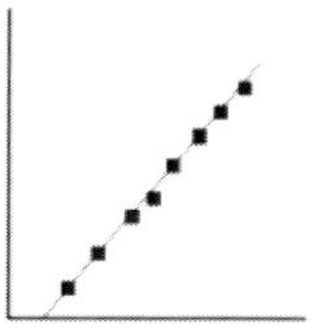

Ⓑ 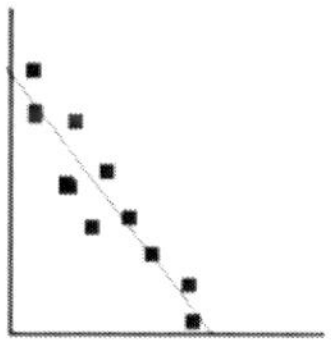

Ⓒ 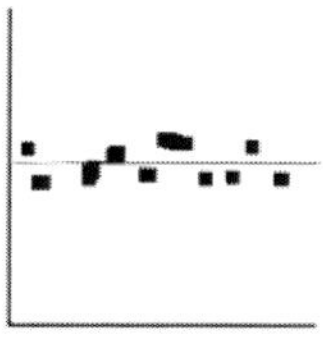

Ⓓ 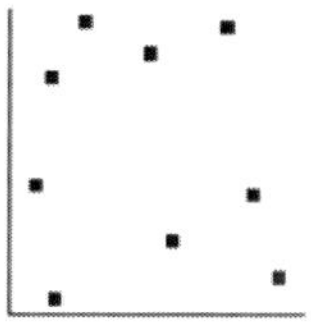

7. **Which scatter plot represents a constant association?**

Ⓐ
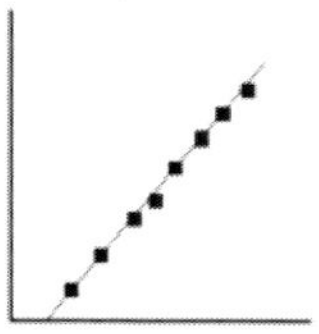

Ⓑ
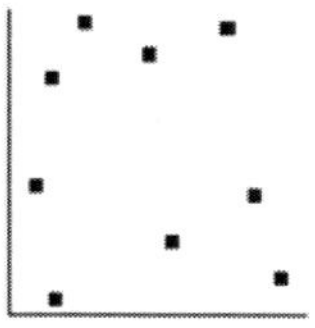

Ⓒ
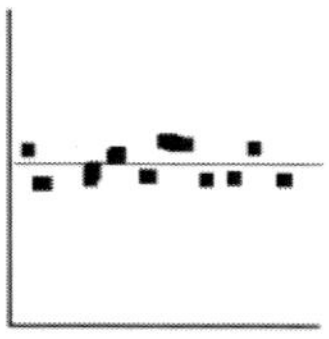

Ⓓ
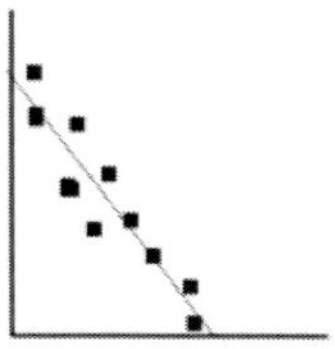

8. **The graph of this data set would most resemble which of the following graphs?**

x	1	2	3	4	5	6	7
y	2	3	4	5	6	7	8

Ⓐ
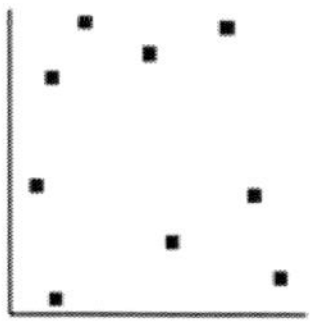

Ⓑ
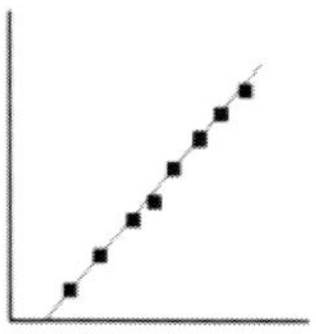

Ⓒ
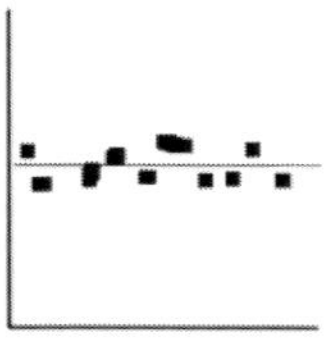

Ⓓ
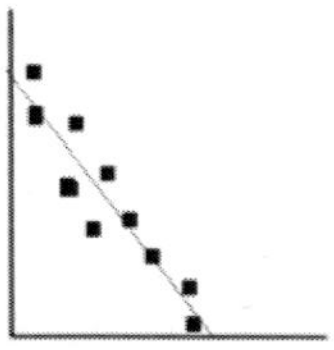

9. The graph of this data would most resemble which of the following graphs?

x	1	2	3	4	5	6	7
y	4	4	4	4	4	4	4

Ⓐ 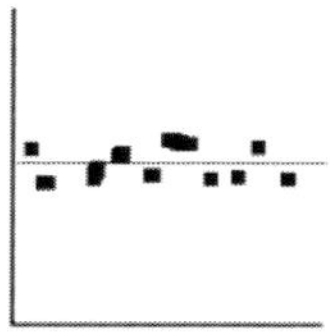

Ⓑ 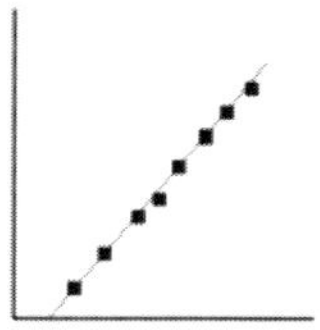

Ⓒ 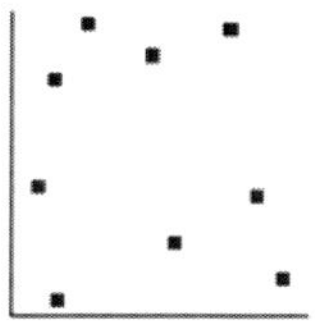

Ⓓ 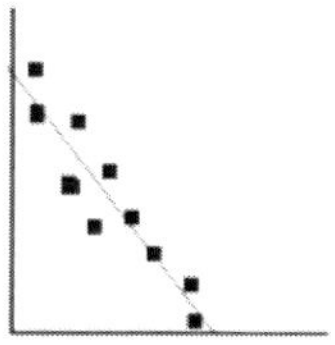

10. Typically air temperature decreases through the night between midnight and 6:00 am. This is an example of what type of association?

Ⓐ constant association
Ⓑ positive linear association
Ⓒ negative linear association
Ⓓ no association

11. Match which line would be the best fit to describe the data pictured.

Figure - 1

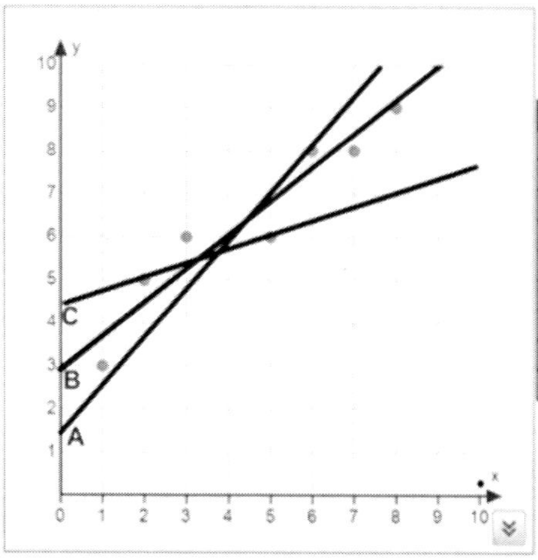

X	Y
1	3
2	5
3	6
5	6
6	8
7	8
8	9
9	10
10	11

Figure - 2

X	Y
1	9
2	8
3	6
5	6
6	5
7	4
8	4
9	2
10	0

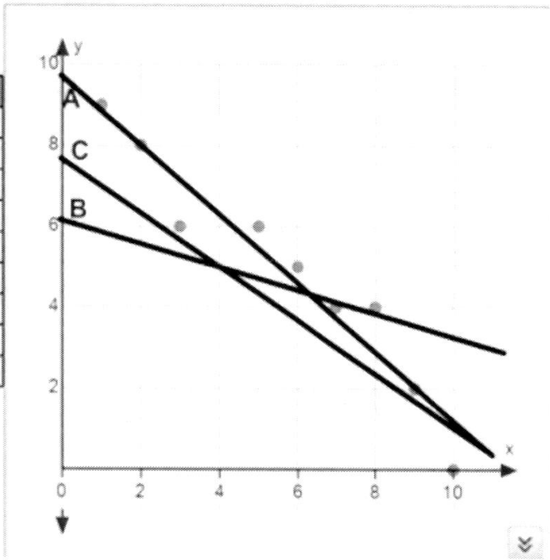

	A	B	C
Figure - 1	○	○	○
Figure - 2	○	○	○

12. Write the prediction equation for this graph using the two labeled points. Leave as fractions.

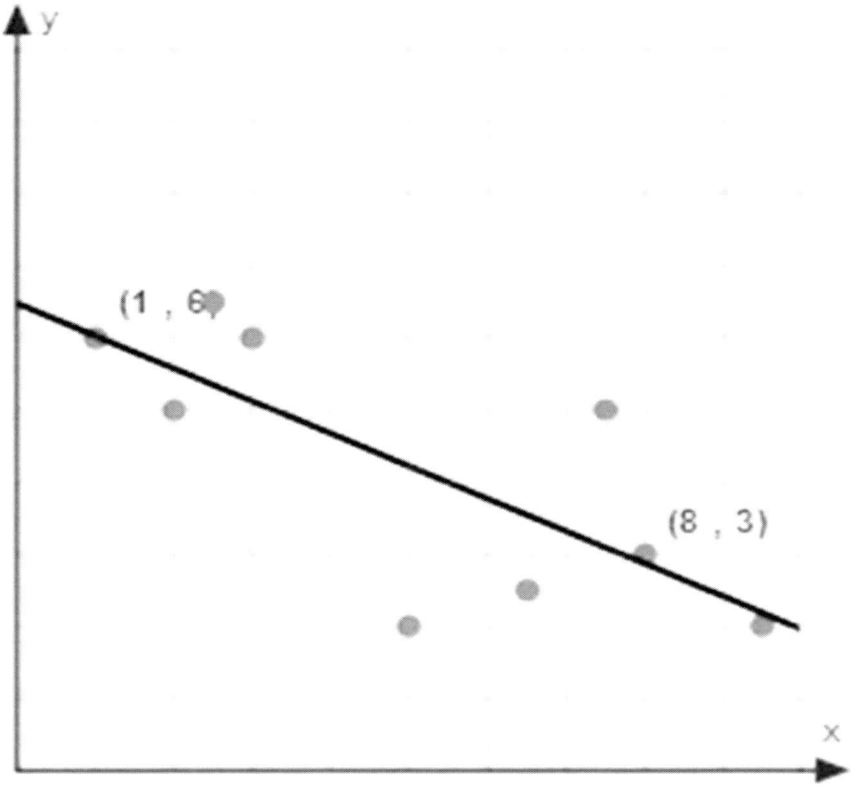

y= ☐ x + ☐

13. Cathy wanted to know what kind of shows the 8th grade class preferred – dramas or comedies. 55 students said they liked comedies and not dramas. 25 students liked both dramas and comedies. There were 41 students who did not like dramas nor comedies. Complete a two way table using the information given.

	Doesn't Like Dramas	Likes Dramas	Total
Doesn't Like Comedies	41		97
Likes Comedies		25	
Total	96		177

CHAPTER 4 → Lesson 3: Sampling a Population

1. Joe and Mary want to calculate the average height of students in their school. Which of the following groups of students would produce the least amount of bias?

Ⓐ Every student in the 8th grade.
Ⓑ Every student on the school basketball team.
Ⓒ A randomly selected group of students in the halls.
Ⓓ Joe & Mary's friends.

2. Which of the following represents who you should survey in a population?

Ⓐ A random, representative group from the population
Ⓑ Every individual in a population
Ⓒ Only those in the population that agree with you
Ⓓ Anyone, including those not in the population

3. What does increasing the sample size of a survey do for the overall results?

Ⓐ Decreases bias in the results
Ⓑ Increases the mean of the results
Ⓒ Increases the reliability of the results
Ⓓ Increasing sample size does not impact the results of a survey

4. Which of the following is not a valid reason for not surveying everyone in a population?

Ⓐ It takes a far longer amount of time to survey everyone.
Ⓑ Not everyone will be willing to participate in the survey.
Ⓒ It is hard to determine the exact size of a population necessary to ensure everyone is surveyed.
Ⓓ Surveying everyone produces unreliable results.

5. Why is it important to know the sample size of a given survey?

Ⓐ It helps determine whether any bias exists.
Ⓑ It helps determine how reliable the results are.
Ⓒ It gives a good estimate for the size of the target population.
Ⓓ It is not important to know the sample size.

6. **John and Maggie want to calculate the average height of students in their school. Which of the following groups of students would most likely produce the most amount of bias?**

Ⓐ Every student in the 8th grade.
Ⓑ Every student on the school basketball team.
Ⓒ A randomly selected group of students in the halls.
Ⓓ John & Maggie's friends.

7. **Which of the following question types will provide the most useful statistical results?**

Ⓐ Open-ended questions where the person surveyed can answer in any way they want
Ⓑ Multiple choice questions offering the person a representative number of choices
Ⓒ True or false questions

8. **Which of the following does not represent a way of avoiding bias in survey results?**

Ⓐ Use neutral words in the questions asked
Ⓑ Ensure a random sample of the population
Ⓒ Only survey individuals that will answer a certain way
Ⓓ Tailor the conclusions based on survey results, not previous thoughts

9. **The following data set represents survey results on a scale of 1 to 10.**

{8, 8, 9, 8, 6, 7, 7, 7, 8, 8, 6}

Which of the following survey result would you be most surprised with if given by the next person surveyed?

Ⓐ 6
Ⓑ 5
Ⓒ 8
Ⓓ 7

10. **The following data set represents survey results on a scale of 1 to 10.**

{6, 6, 7, 6, 8, 7, 7, 7, 6, 6, 8}

Which of the following survey results would you be most surprised with if given by the next person surveyed?

Ⓐ 6
Ⓑ 10
Ⓒ 8
Ⓓ 7

11. **A company is conducting a survey on their performance with customer service. What method will best avoid receiving biased data?**

Ⓐ The company should make the survey anonymous.
Ⓑ The company should take a video survey.
Ⓒ The company should require the survey during the transaction.
Ⓓ The company should just request that everyone fill out a survey.

12. **A company is conducting a survey on their performance with customer service. What should the survey look like?**

Ⓐ The survey should contain a few simple multiple choice questions with an optional comment section.
Ⓑ The survey should contain simple free response questions with an optional comment section.
Ⓒ The survey should contain only free response questions.
Ⓓ The survey should contain some personal questions.

13. **Where should a survey about personal items purchased be conducted?**

Ⓐ The receipt should have a link to the survey website.
Ⓑ The survey should be near the register.
Ⓒ The survey should be near the exit doors.
Ⓓ The survey should be in the parking lot.

14. **How should online video game surveys be conducted?**

Ⓐ This survey should be put at the end of a level or section of the game.
Ⓑ This survey should be put at the beginning of the game.
Ⓒ Thus survey should be put on flyers and distributed at a gaming store.
Ⓓ This survey should be put on the website separate from the game.

15. **A large corporation is launching a new product, Hitz, which allows customers to purchase music and store it on the corporation's servers. They want to survey people who listen to a lot of music. Which of the following would give them the best sample?**

Ⓐ Conducting a survey inside of a music store
Ⓑ Conducting a survey at the parking lot of a music concert
Ⓒ Conducting an online survey on music social networking websites
Ⓓ Conducting a survey of students in a school marching band.

16. Match the situation with the sampling method used.

	Convenience Sample	Systematic Sample	Simple Random Sample
A person chooses every 5th person on a list of names starting with #1.	○	○	○
A person accepts the first 15 people to respond to a magazine ad.	○	○	○
A person picks names out of a hat.	○	○	○

17. A jar of marbles contains gray and black marbles. You collect the representative sample shown here:

If the jar contains 50 marbles, about how many marbles are gray?

Write your answer in the box given below.

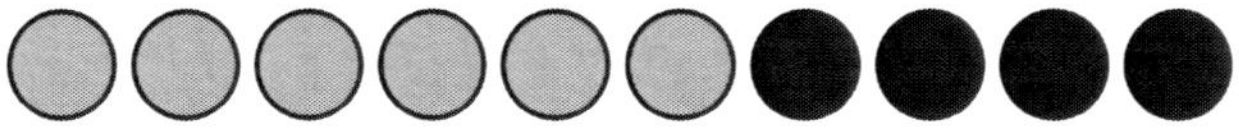

ANSWER KEY AND DETAILED EXPLANATION

Chapter 4 Data & Probability

Lesson 1: Interpreting Data Tables & Scatter Plots

Question No.	Answer	Detailed Explanation
1	B	By definition, a decreasing trend from left to right on a scatter plot indicates a negative association.
2	A	By definition, an increasing line from left to right on a scatter plot indicates a positive association.
3	C	The points in the third choice are nearly in a straight line.
4	A	As the History grade increases, so does the English grade. Thus, there is a positive association.
5	A	There does not appear to be any significant correlation between these two variables.
6	A	As the Science grade increases, so does the Math grade. This indicates a positive association.
7	A	In the first graph, a best fit line would be nearly a straight line increasing to the right.
8	A	$30\times\frac{100}{150} = 20\%$
9	B	Option (B) is the correct answer. 80 do not play an instrument and 50 of those do not play a sport. Among these 80 students, ratio of those who do no play a sport to those who play a sport is 50:30 = 5:3, It is not 2:1 as given in the option (B). All the other statements are supported by the data.
10	A	30+50=80, which is about half of 150.

Question No.	Answer	Detailed Explanation
11		If there is a positive association, as one item increases so does the other. If there is a negative association, as one increases the other decreases. The last one is no association. This happens if there is no pattern as to what is happening.
12	no association	There is no association between the apple and mango sales. You can see that they neither consistently go up or go down over the course of the 10 days.
13	negative|| none|| positive|| negative	The first graph has a negative association because as you move across the graph you can see that the points plotted continually decrease. The second graph has none because there is no pattern to the points. The third graph has a positive association because as you move across the graph you can see that the points continually increase. The last graph will also be a negative association because as you move across the graph you can see that the point plotted continually decrease.

Lesson 2: Scatter Plots, Line of Best Fit

Question No.	Answer	Detailed Explanation
1	D	This scatter plot does not represent a linear function. None of these is the correct choice.
2	C	The line of best fit would be a horizontal line. Constant linear is the correct choice.
3	D	The data does not represent a straight line. None of these is the correct choice.
4	A	The first scatter plot shows a positive slope representing a positive linear association.
5	B	The second scatter plot shows a negative slope representing a negative linear association.
6	D	The data on the fourth scatter plot cannot be represented linearly and, therefore, represents no association.
7	C	The third scatter plot represents a horizontal line or a constant association.
8	B	The data represents a straight line with a slope of positive 1. The second scatter plot is the correct choice.
9	A	The y-coordinate is a constant 4 representing a horizontal line at y=4. The first scatter plot is the correct choice.
10	C	Temperature decreases as the hours increase. Negative linear association is the correct choice.
11		To get the best fit line for the data, it won't necessarily go through every point, but will show the general trend of the data.
12	y=(-3/7) x+45/7	The best fit line is already drawn on the graph, but they want the equation to go with it. Using the 2 points they gave us we first need to find the slope. $\frac{6-8}{1-8} = -\frac{3}{7}$ Now that we know the slope, we can fill in an ordered pair and solve for the y-intercept. $6=-\frac{3}{7}(1)+b$ When you solve for b you get . Put those numbers together and you get the equation . $y=-\frac{3}{7}x+\frac{45}{7}$

<table>
<tr><th>Question No.</th><th>Answer</th><th>Detailed Explanation</th></tr>
<tr><td>13</td><td>(1) 56
(2) 55, 80
(3) 81</td><td>

	Doesn't Like Dramas	Likes Dramas	Total
Doesn't Like Comedies	41	**56**	97
Likes Comedies	**55**	25	**80**
Total	96	**81**	177

To solve how many like dramas and doesn't like comedies, you could take 97 - 41 to get 56. Next you know that 55 liked comedies but not dramas. To get the total number of students who liked comedies, just add 55 + 25 to get 80. Lastly to get the total of how many like dramas, take 56 + 25 to get 81.</td></tr>
</table>

Lesson 3: Sampling a Population

Question No.	Answer	Detailed Explanations
1	C	A group of randomly selected students in the hallways would produce the least amount of bias because it is unlikely for assumptions to be made or factors that influence the data to be present. For example, students in the 8th grade may be taller than other students in other grades, skewing the data toward a higher average. A similar assumption can be made about students on the basketball team. Joe and Mary would already have an idea of the height of their friends.
2	A	A random, representative group represents the people to survey in a population because each person in the population has an equal chance of being included and there is less of a chance of bias altering the results of the survey.
3	C	As the size of the sample (people surveyed) increases, the results become more accurate. Therefore, increasing the sample size increases the reliability of the results.
4	D	Surveying everyone would produce reliable results because there would be specific data from each person; however, due to the length of time needed, among other reasons, surveying everyone may not be possible (depending on the reasons for the survey).
5	B	A sample size gets information from a small group from the population. Since the sample size is supposed to represent the population at large, it's important to know the parameters of the sample size in order to determine how reliable the results are for a given survey.
6	B	An assumption can be made that students on the basketball team are likely to be taller than other students at the school, which can influence the results (make it appear that the average height of the general school population is higher than it actually is). Therefore, calculating the average height of the basketball team would most likely produce the most amount of bias.
7	B	Multiple choice questions are useful for statistical results because they can represents different amounts of data that can be separated into categories and sub-categories based on the type of survey and the intended outcome.

Question No.	Answer	Detailed Explanations
8	C	Bias in collecting data can be caused by using questions where there are likely to be assumptions made or factors that influence the data. A survey that asks questions that cause individuals to answer in a certain way makes assumptions and influences the outcome of the data. Therefore, using that type of survey does not avoid bias.
9	B	When analyzing the data, out of the 10 responses, none of the responses have been less than 6. Therefore, a surprising response from the next person would be a 5 since it is outside the range.
10	B	When analyzing the data, out of the 10 responses, all are within the range of 6 to 8. Therefore, a surprising response from the next person would be a 10 since it is outside of the range.
11	A	Bias in collecting data can be caused by using questions where there are likely to be assumptions made or factors that influence the data. An anonymous survey would encourage individuals to answer questions openly without worry about the implications of their answers. Therefore, an anonymous survey is the best way to avoid receiving biased data.
12	A	Multiple choice questions are useful for statistical results because they can represent different amounts of data that can be separated into categories and sub-categories based on the type of survey and the intended outcome. The optional comments section gives an individual the opportunity to make open observations (without bias). A survey with too many free response questions would be hard to analyze and personal questions invite bias.
13	A	For collecting personal data, an anonymous survey would encourage individuals to answer questions openly without worry about the implications of their answers. It is also a way to avoid receiving bias data. Therefore, a receipt should have a link to an online survey where individuals can enter information about personal data.
14	A	In order to get a reliable, random sample based on the population (players who use the game), the surveyors should set up their survey at the end of a level or section of the game. Players are more likely to answer the survey before going to the next level. People are ready to play at the beginning of a game and ready to turn the game off near the end, so there is a good chance they won't complete the survey. A separate survey is likely to be ignored.

<table>
<tr><th>Question No.</th><th>Answer</th><th>Detailed Explanations</th></tr>
<tr><td>15</td><td>C</td><td>When choosing a sample to survey for data, the sample should be representative of the target population. Here, the target population is people most likely to purchase music online. These people are most likely to communicate with others about music online. Therefore, an online survey on music social networking websites would give them the best sample.</td></tr>
<tr><td>16</td><td colspan="2">
<table>
<tr><th></th><th>Convenience Sample</th><th>Systematic Sample</th><th>Simple Random Sample</th></tr>
<tr><td>A person chooses every 5th person on a list of names starting with #1.</td><td>○</td><td>●</td><td>○</td></tr>
<tr><td>A person accepts the first 15 people to respond to a magazine ad.</td><td>●</td><td>○</td><td>○</td></tr>
<tr><td>A person picks names out of a hat.</td><td>○</td><td>○</td><td>●</td></tr>
</table>
Systematic sampling is a type of probability sampling method in which sample members from a larger population are selected according to a random starting point and a fixed, periodic interval.
Convenience sampling is a statistical method of drawing representative data by selecting people because of the ease of their volunteering or selecting units because of their availability or easy access.
A simple random sample is a subset of a statistical population in which each member of the subset has an equal probability of being chosen. A simple random sample is meant to be an unbiased representation of a group.
</td></tr>
<tr><td>17</td><td>30</td><td>If the representative sample is 10 then you would multiply the number of gray marbles by 5 to get the approximate number of gray marbles. 6 x 5 = 30.</td></tr>
</table>

Additional Information

Test Taking Tips

1) **The day before the test,** make sure you get a good night's sleep.

2) **On the day of the test,** be sure to eat a good hearty breakfast! Also, be sure to arrive at school on time.

3) **During the test:**

- **Read every question carefully.**
    - Do not spend too much time on any one question. Work steadily through all questions in the section.
    - Attempt all of the questions even if you are not sure of some answers.
    - If you run into a difficult question, eliminate as many choices as you can and then pick the best one from the remaining choices. Intelligent guessing will help you increase your score.
    - Also, mark the question so that if you have extra time, you can return to it after you reach the end of the section.
    - Some questions may refer to a graph, chart, or other kind of picture. Carefully review the graphic before answering the question.
    - Be sure to include explanations for your written responses and show all work.
- **While Answering Multiple-Choice (EBSR) questions.**
    - Select the bubble corresponding to your answer choice.
    - Read **all** of the answer choices, even if think you have found the correct answer.
- **While Answering TECR questions.**
    - Read the directions of each question. Some might ask you to drag something, others to select, and still others to highlight. Follow all instructions of the question (or questions if it is in multiple parts)

Frequently Asked Questions(FAQs)

For more information on the assessment, visit
www.lumoslearning.com/a/ostp-faqs
OR Scan the **QR Code**

What if I buy more than one Lumos tedBook?

Step 1 → Visit the link given below and login to your parent/teacher account

www.lumoslearning.com

Step 2 → <u>**For Parent**</u>

Click on the horizontal lines (≡) in the top right-hand corner and select **"My tedBooks".** Place the Book Access Code and submit.

<u>**For Teacher**</u>

Click on "My Subscription" under the "My Account" menu in the left-hand side and select **"My tedBooks".** Place the Book Access Code and submit.

Note: See the first page for access code.

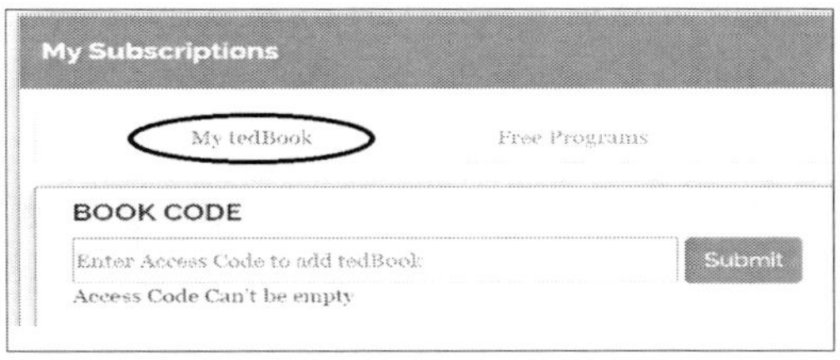

Step 3 → Add the new book

To add the new book for a registered student, choose the '**Student**' button and click on submit.

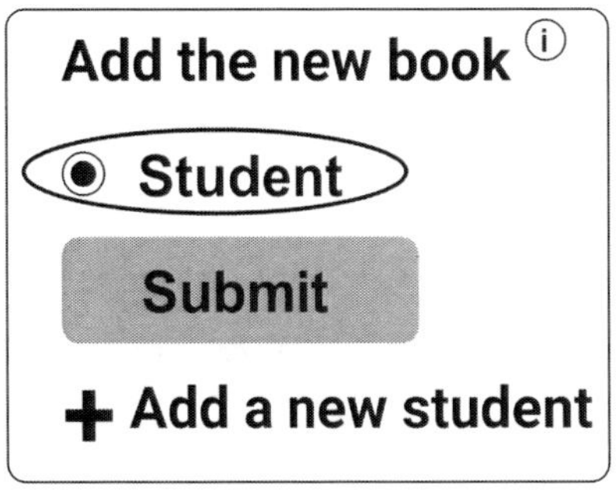

To add the new book for a new student, choose the '**Add New Student**' button and complete the student registration.

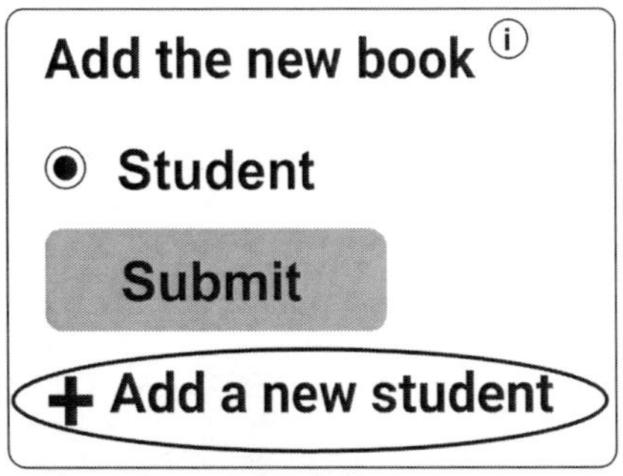

Progress Chart

Standard	Lesson	Score	Date of Completion
OAS			
PA.N.1.1	Properties of Exponents		
PA.N.1.2	Scientific Notation		
PA.N.1.3	Solving Problems Involving Scientific Notation		
PA.N.1.4	Square and Cube Roots		
PA.A.1.1	Functions		
PA.A.1.2	Linear Function Models		
PA.A.1.3	Linear Functions		
PA.A.2.1	Comparing Functions		
PA.A.2.2	Analyzing Functions		
PA.A.2.3	Analyzing Linear Scatter Plots		
PA.A.4.1	Solving Linear Equations		
PA.A.4.3	Solve Linear Equations with Rational Numbers		
PA.GM.1.1	Verifying the Pythagorean Theorem		
PA.GM.1.1	Pythagorean Theorem in Real-World Problems		
PA.GM.1.2	Pythagorean Theorem and Coordinate System		
PA.GM.2.4	Finding Volume: Cone, Cylinder, and Sphere		
PA.D.1.2	Interpreting Data Tables and Scatter Plots		
PA.D.1.3	Scatter Plots, Line of Best Fit		
PA.D.2.2	Sampling a Population		

Made in United States
North Haven, CT
24 February 2026

89101519R00072